T0133725

Essential Computer Graphics Techniques for Modeling, Animating, and Rendering Biomolecules and Cells

A Guide for the Scientist and Artist

Essential Computer Graphics Techniques for Modeling, Animating, and Rendering Biomolecules and Cells

A Guide for the Scientist and Artist

By Giorgio Luciano

CRC Press
Taylor & Francis Group
Boca Raton London New York

CRC Press is an imprint of the
Taylor & Francis Group, an **informa** business
AN A K PETERS BOOK

CRC Press
Taylor & Francis Group
6000 Broken Sound Parkway NW, Suite 300
Boca Raton, FL 33487-2742

Library of Congress Cataloging-in-Publication Data

Names: Luciano, Giorgio, author.
Title: Essential computer graphics techniques for modeling, animating, and rendering biomolecules and cells : a guide for the scientist and artist / Giorgio Luciano.
Description: Boca Raton : Taylor & Francis, a CRC title, part of the Taylor & Francis imprint, a member of the Taylor & Francis Group, the academic division of T&F Informa, plc, 2019. | Includes bibliographical references.
Identifiers: LCCN 2018045164 | ISBN 9781498799218 (hardback : acid-free paper)
Subjects: LCSH: Computer graphics.
Classification: LCC T385 .L8335 2019 | DDC 571.60285/66–dc23
LC record available at https://lccn.loc.gov/2018045164

Visit the Taylor & Francis Web site at
http://www.taylorandfrancis.com

and the CRC Press Web site at
http://www.crcpress.com

Contents

Acknowledgments, vii

CHAPTER 1 ▪ Preface   1

CHAPTER 2 ▪ Introduction   9

CHAPTER 3 ▪ Foundations   13

CHAPTER 4 ▪ Modeling and Lighting   29

CHAPTER 5 ▪ Scene Setup   75

CHAPTER 6 ▪ Rendering   97

CHAPTER 7 ▪ Animation   149

CHAPTER 8 ▪ Final Look   183

CHAPTER 9 ▪ Professional Practices   203

INDEX, 211

Acknowledgments

The fantastic staff of Taylor and Francis; Rick, for giving me the possibility to write this book; Jessica for helping me keep the pace and the useful advice given during the writing; Joette and Ananth for their patience and help during the production stages.

All the friends at CGsociety that first introduced me to the world of computer graphics and in particular Kirsty Parkin for believing in me at first and Zoe Watson for the patience while helping me in the first workshops. All the people at CGMA; Ted, Rachel, Josh for keeping the workshops going and helping at every step of them. Travis at CGTalk and IMT for his patience and support.

Miriam Payne, Anette Martinez and David Diez at Foundry. Julie Lottering from sideFX. Jorge Bazaco, Ben Fischler, Rebecca Purver, and Alexandra Constantine from Solidangle/Autodesk and Andres Hildebrandt from Maxon for giving me the opportunity to test and use their fantastic software.

My colleagues at the Institute for the Research of Macromolecules for being an incredible team to work with.

My grandpas and grandmas for being a source of inspiration. My parents and sister for everything (no words needed).

CHAPTER 1

Preface

During 2013, while I was forum leader for the lighting challenge at the CGtalk forum, Kirsty Parkin was looking for ideas for new workshops/courses at the CGSociety. I read her message while I was at work at the National Research Council, and then the cover of a scientific journal got my attention. It was a very high impact journal devoted to a wide range of scientific disciplines. Although the concept on the cover was clear and sound, the 3D graphics looked cliché and dated, so I sent an email to Kristin with this subject: "What do you think about creating a workshop about scientific visualization?" My aim, at that time, was not too ambitious. I would have been very happy just to be able to teach the basics of 3D to researchers, since I had already taught computer graphics courses for my colleagues, and I was very satisfied with the results of just a few days of teaching. It was very funny to open a small world for the scientists involved, letting them know the basic concepts of 3D modeling and shading and seeing how they enjoyed creating their first image (the mandatory teapot with a shader applied) and a basic three lighting setup in all its glory! It was also a way to exchange scientific ideas with other people; it was enriching to understand their approach to this subject after the more artistic approach to which I was accustomed from moderating threads and giving feedback at CGTalk.

Kirsty thought that the idea of teaching a course in biochemical visualization (actually, we were quite unsure what to call the workshop) was interesting since it was a field that was growing, testified by the fact that they had already created a CGChallenge about the visualization of the HIV virus. She was interested in the subject, but it was not easy to find people versed in both science and CG. So, thanks to Kirsty, I started teaching the first Biomedical Illustration Workshop at the CGSociety. I admit that I was very anxious and happy at the same time since I was not sure what to expect.

The first challenge I had to face was a big dichotomy among my students. Half of them were scientists with almost no experience in 3D; the other half was made up of professionals working in the CG industry (mainly motion graphics and advertising) who wanted to know what skills were needed to create their illustrations (still images, small animations, previz, movies, etc.).

I went for a customized approach for each student, trying to focus on his or her strong and weak points, suggesting where and how to improve with the help of correct references and exercises of increasing difficulty.

I soon realized that, even if the tools evolved quickly, the work of an illustrator is to create bridges between scientists and an audience in an effort to represent concepts as accurate, plausible, and at the same time as beautiful and artistic as possible. The caveat is what every translator faces in his or her work. In Italy, my country, there is a phrase *traduttore traditore*, translated literally as translator traitor. In the effort to represent concepts in an artistic way and to make them approachable for a wider audience, we should never forget the rigor of what we are depicting. This book represents a small effort to help illustrators learn new tools and scientists become better acquainted with the basic concepts of illustration.

Science has made giant steps in the last years toward making available data on the structure of molecules, cells, human and animal anatomy, etc. Often, the beauty of reality oversteps our artistic vision. The more information and tools we have on the concepts we are going to illustrate, the better. In this book, I will give information on where to find it and tools that can be helpful while representing scientific concepts.

Knowing the basics of 3D software will also improve the way we understand concepts and create mental models since it will force us to have a clear vision of what we want to communicate to the world.

FOR WHOM IS THIS BOOK INTENDED?

1. Scientists and researchers that would like to understand the basics of animation/lighting and rendering in 3D graphics

2. Artists who would like to know more about the scientific software used in the biochemical/medical field

3. Students of CG who would like to know more about biochemistry/biomedical rendering/animation and to look at the software used in the biomedical field

4. Teachers who would like to represent in a catchy way the concepts of the biochemical/medical field and want to learn the basics of illustration software

Special attention will be dedicated to the Edutainment world. The main effort of the book is to merge the use of scientific software and mainstream 3D packages in a workflow to produce artistic and pleasant images with strong scientific roots. If you are a 2D/3D artist, this is NOT a book for becoming a scientist in the biomedical/biomolecular field, even though it will show you fascinating concepts and ideas in these fields. It is, rather, a starting point for gathering references for creating your own path toward working with scientists to create a common background.

Style of the Book

The book is informal, despite my best effort to be rigorous while describing scientific concepts. I wrote every chapter to be self-sufficient and to encourage the reader to explore his or her own topic of choice and create a personalized learning path. For this reason, I preferred to include a *selected* list of references at the end of each chapter. I listed only what I think are the indispensable and mandatory references in order not to overwhelm the reader, considering the increasing quantity of available literature on each topic

Most chapters include

- Keywords

- A chapter outline

- Focused sessions on specific topics that describe special skills and tools for achieving a specific rendition of the concepts presented

- Tutorials on the key illustrations presented in each chapter

- Exercises to help the reader build his or her knowledge on the subject

- Icons of the software used to create specific tutorials and images in order to help the reader choose those most appropriate

WHICH SOFTWARE?

I started to "play" with 3D software in the '90s when I was a college student. The first program I used was Autodesk 3D Studio Max. I like to learn new skills, so during the years I've tried Maya, Cinema4D, and Pixologic Zbrush; I've also started digging into Modo and Houdini (hard to learn but unrivaled in the field of generative and procedural modeling). Now Zbrush, Cinema4D, and Modo are indispensable tools of my everyday job.

I have tried Blender, and I'm very happy that I did. It is a pro tool with tons of add-ons and is Python scriptable. Since I have to use Python quite often to manipulate my data, I feel at home and use it to create examples of scripts for procedural modeling.

When I teach at Computer Graphics Master Academy (CGMA), every student has a different background in 3D and is skilled in specific software. The first question everyone generally asks is: May I attend your course if I use Software X? The second most asked question is: Is Software X better for biomedical illustration than Software Y? Is there a software that has all the features we need?

The answer is: "Of course not, otherwise we all would use THAT software." Everyone has personal preferences and different pockets. It would be a huge task to show how to create illustrations with EVERY 3D software, and it can be self-limiting to try to accomplish everything with just ONE software. In this book, I will focus on the similarities among software I use, and I will teach the general concepts behind all of them. In this way, you will be more flexible and it will not be too complicated to shift the tools for your work depending on the preference of the studio where you work or the development of new software.

When we work on biochemical/biomedical medical representation, we will need to use a lot of very specific packages used in everyday work by scientists. Luckily, a lot of software in the technical field is open source, free, and available for several operating systems.

In the (rare) case that there is a task that can only be accomplished with one specific software, I will highlight it.

My Workflow

This section relates to the process of creating a product for a wide audience, not a product for peers. I start almost every project with paper and pencil. I write down a few lines and create a very primitive storyboard of how the animation should look. In this stage, I prefer pen and paper to software, since it gives me total freedom to doodle. Since it is a kind of "creative" project, I need to feel free. The second stage is about *modeling*. First, I try to find as many *references* as possible. If the animation is about organic molecules/nanomachineries, I look at the protein data bank (PDB) and eventually download the files needed. Then I search the EMDB database to see if there are other data available with techniques not present on the PDB site. After this step, it is time to import the coordinates of the molecules into the dedicated software: Chimera and VMD. I look at the molecules, check for a suitable representation among the several available, and eventually create a small animation. Then I export the animated model into a format that can be imported in my main 3D application (generally alembic format abc or dae).

If I need to represent inorganic complexes and molecules, I also use Vesta and Nanoengineer Nanorex, dedicated to this branch of chemistry.

I refine the model in Zbrush and export it in order to use it in Cinema4D, Modo, and/or Houdini. Cinema4D is quite an easy software, and I prefer it for easy rendering, Modo is my battle horse, since it can do almost anything. Houdini is my companion for special effects and animation (it can be used for modular modeling, as we will see later in the book). I set up *lightning and then shaders* (textures and the properties of the molecules/objects represented) with the help of specific software such as Substance Designer. Then I animate the models, give a fast try at rendering using low settings, and then start again and set up the details in the scene. After the animation is ready, I render all the frames and export them in an uncompressed image file (TIF or TGA). My favorite render engines are Arnold, Renderman, and Mantra. I composite my passes using Blackmagic Fusion, Nuke, or Natron.

A few key points:

- Put the ideas on paper using traditional techniques or very easy tools so as not to focus on the tools (pencil and paper or a quick sketching application)

- Search for references: Probably the most important part of the whole process. Never get tired of doing it; every minute you spend searching reference will be well spent.

- Create/Import your model.

- Create mockup animation if needed.

- Light your scene.

- Apply shaders and texture (the appearance of your model), and recheck your lighting.

- Render the animation (low resolution to check if everything is in place).

- Recheck everything and repeat the necessary steps.

ADVICE ABOUT FILE NAMING AND ORGANIZING YOUR FILES

KEEP YOUR FILES CLEAN! (uppercase for emphasis intended). Keeping your files organized is one good habit that will save you tons of time, in comparison to keeping everything in unique folders with names like untitled01.3ds, first.ob, 01, second.obj, final_one.bck, really_final_one.fbx etc.

Even if it seems obvious and quite easy to do, its complexity can escalate quickly depending on the size of your project. A very good, generic introduction that will help you keep your non-CG files in a clean and organized structure is included in the book, *Guerrilla*

Analytics: A Practical Approach to Working with Data. If your everyday job includes using a computer (so every job), I strongly encourage you to read it. It is really worth your time. After reading several threads on this topic on different pro forums (CGSociety, Polycount, etc.), and with the help of Hilaire Gagne (an animation pro), I choose to organize my files using this tree

```
-Assets
-Lighting Test
-Models
-fbx
-obj
-abc
-References
-Shaders
-Textures
-non pbr
-pbr
```

In this way, you should have everything at hand and find it quite quickly.

Since the chaotic me sometimes needs to take a pause from the company of the OCD me, I created a folder called daily_box. The rule is that I can put everything I want there, but at the end of the day, I MUST clean it out.

Remember that every studio has its own **very strict** convention for naming. This is not only because of OCDs but also because scripts need to find the files with their own exact names.

The convention I prefer to use is to put underscores (since not all OSs like spaces) and a small description of the file, for example

```
Body3D_anatomy_v1_remeshed
BloodCells_vein_animation_v1_no_texture
```

If you succeed in sticking to it, this will greatly help in your everyday work. If not, there is always the UNSORTED folder as a last resort.

Software and Hardware

Every scene presented in the book was tested with the following configuration.

Software: Modo 10.2v4, Cinema4D R18, Houdini 16.0.2.54, Arnold Render 4.2, Substance Designer 6.0, Substance Painter 2.5. Chimera 1.01, Vmd 1.56, Vesta 3.0, Avogadro 2.0, ImageJ 2, Invesalius 3.0, EPMV 2.43.

Hardware Configuration: Intel I3 4660, 32 Gb Ram, 256Gb SSD, 2TB Raid Sata 3, nVidia GTX 1050Ti.

DOWNLOADING ADDITIONAL FILES

Throughout the book, you will find references to scenes available for download at the author's book website. Feel free to experiment with them for your personal work and share your results.

In order to ensure future compatibility, all assets (light rigs, models, and animations), apart from being made available in their native format, were translated in alembic (a de facto standard in 3D graphics and animation).

CHAPTER 2

Introduction

REPRESENTATIONS IN CHEMISTRY

In chemistry, we have two kind of visualization: *concept visualization* and *data visualization*. Molecular structure and dynamics belong to the first category, while graphs and renderings from instrumental data such as images from an electronic microscope. The most famous representation in chemistry is, without doubt, the periodic table of the elements, a two-dimensional representation for data that need at least three dimensions to clearly represent relationships.

Concept visualization has helped chemists in problem solving since the early days of chemistry. Kekule's discovery of benzene and Mendeleev's discovery of the periodic organization of the elements, even if they are apocryphal, depict the process of solving problems by means of mental images.

Concept visualizations are metaphors, symbols, and signs that in a chemist's mind are reconstructed in three-dimensional structures. They are the part of a chemical language that despite its imprecision help people communicate and bring richness and ambiguity.

Our metaphors are plastic and can serve to solve a wide range of problems. How can we represent objects, those small bricks smaller than visible light, that carry unique features we need to highlight? We have a wide toolbox containing ribbons and chains of different thickness for different kind of bonds, but let's start from the beginning.

More than 100 years ago, Dalton made a systematic series for the visualization of atoms and started to differentiate elements using symbols from alchemy.

Hofmann, an Austrian chemist who studied as an architect, invented the ball and stick models in the 19th century. Van 't Hoff and Le Bel introduced three-dimensional models that for the first time focused on the depiction of organic molecules. Stuart in the '30s was the first to create models that took into account the volumes of the atoms; in the '60s, Corey-Pauling-Koltum (CPK models) were born and then subsequently modified to introduce rotation in bonds.

Then we have a quantum leap with the introduction of the computer. We started to generate models from experimental techniques. We could almost instantly redraw models to highlight different functions of different parts of the molecule (i.e., chains in a protein) and depict different properties like the connectivity of the atoms, shapes, charges, etc.

In the 1980s, a new technology was introduced: the scanning tunneling microscope (STM), with the help of computers, allowed for the depiction of molecules from experimental data. For the first time, it was possible to check whether some theoretical depictions used had real counterparts.

From the first computer static representations, we gradually evolved to animations that can involve thousands of atoms; we can study molecule bonds and active sites in macromolecules, visualize polymerization processes, and so on.

We can see the effects of this evolution in textbook visualization, from the first texts in the '50s, when creating and depicting concepts was still expensive, to the pioneering work of Geis (also an illustrator) and Atkins. In the '70s we started to see a few colored pages, and in the '80s, color was an integral part of textbooks (this is also thanks to the revolution in the printing process where chemistry has its own merits). Currently, textbooks include multimedia contents (i.e., animations and 3D representations to download, increasing exponentially the field of the work of illustrators).

The same revolution is true for scientific journals, where ubiquitous linking to electronic versions help in the picturing of concepts whose entire complexity cannot be rendered on printed paper.

The representation of chemical concepts is an extremely exciting field, more alive than it ever was and present in all the media including emerging technologies. We hope that virtual and augmented reality survive their second coming and the hype they are creating; the possibility to navigate into cells at the atomic level has become a reality, and we need to refine the metaphors we use in representation to get the most from what they have to offer.

FROM ATOMS TO HUMANS, A LEAP OF FAITH

As reported in an informative minireview by David Goodsell, published in *Structure* (2005, p. 353)

> Molecular biology is gradually merging with cell biology as data become increasingly available over the entire scale range from nanometers to millimeters.

It is useful to represent in a consistent hierarchical style across scale all of the information we have about molecules/nanomachineries. The metaphor we use in this case is "enlarging" our objects to a visible size. We smooth our representation the larger they become in the aim to keep our representation readable while keeping the complexity of the details represented manageable.

We don't have (yet) a *unique* technique that can help us investigate from atomic to tissue and cell structure, but we can fill in the gaps in a plausible way. The same is true concerning human anatomy, but in this case we have a multitude of data, gathered from several different techniques that can help us in a way it was impossible just a few decades ago. Thanks to physics and its contributions to imaging techniques in the medical field, we can have the accurate representation of organs for every single patient and representation in real time of our living body. We are starting to merge this information using virtual reality and augmented reality techniques mentioned above, realizing what we only dreamed just few years ago. These techniques keep improving our knowledge of the human body and create new useful metaphors; just think about the "discovery" of a new organ in our body in 2017.

SUGGESTED READINGS

Goodsell, David S. 2005. "Visual Methods from Atoms to Cells." *Structure* 13 (3): 347–354. doi:10.1016/j.str.2005.01.012.

Habraken, Clarisse L. 1996. "Perceptions of Chemistry: Why Is the Common Perception of Chemistry, the Most Visual of Sciences, so Distorted?" *Journal of Science Education and Technology* 5 (3): 193–201. doi:10.1007/BF01575303.

Hoffman, John M., and Sanjiv S. Gambhir. 2007. "Molecular Imaging: The Vision and Opportunity for Radiology in the Future." *Radiology* 244 (1): 39–47. doi:10.1148/radiol.2441060773.

Hoffmann, R. (2009). "Abstract Science?" *American Scientist.* doi:10.1511/2009.81.450

Johnson, Graham T., and Samuel Hertig. 2014. "A Guide to the Visual Analysis and Communication of Biomolecular Structural Data." *Nature Reviews. Molecular Cell Biology* 15 (10). Nature Publishing Group: 690–698. doi:10.1038/nrm3874.

Le Muzic, M., L. Autin, J. Parulek, and I. Viola. 2015. "cellVIEW: a Tool for Illustrative and Multi-Scale Rendering of Large Biomolecular Datasets." *Eurographics Workshop on Visual Computing for Biomedicine.* doi:10.2312/vcbm.20151209

Sali, Andrej, and Chiu Wah. 2005. "Macromolecular Assemblies Highlighted." *Structure* 13 (3): 339–341. doi:10.1016/j.str.2005.02.010.

CHAPTER 3

Foundations

E very time we look at the model of a molecule in a book or an article, we are looking at a visual metaphor, since it is not possible *to take a picture* at atomic scale. The objects we would like to observe are smaller than the visible wavelength, so we need to generate synthetic images. We have several tools to help us in this task, and each one helps us focus on a different aspect that we would like to communicate to the public.

The first visualization category is *data analysis*. It is an abstract form of science communication. Think about a plot created with dedicated software (i.e., Avogadro or Gabedit) in researching for a scientific paper. We need an immediate and interactive visualization of our data. Everything needs to be essential; superfluous information would only impair our intended focus on and elaboration of results.

We use a second representation in communicating results to an "educated" audience; we can call this *peer communication*. Think about the diagrams used at a presentation when we have to rigorously communicate our results to someone working in the same field as ours (e.g., the illustration of the synthesis of a novel drug and its steps, a new technique for gene editing, the physicochemical characteristic of a material).

Another way of communication is necessary when dealing with *educational resources*. Books are the older example in this field. It needs to be rigorous, more captivating in comparison with peer communication, and needs to highlight the essential step we would like our students to remember (e.g., the classic representation of citric acid cycle). Today we can reach beyond the limits of printed paper and take full advantage of all that technology can offer: movies directly recorded from instruments while performing analysis, animations, educational software and applications, augmented and virtual reality, everything that will help us describe the mechanisms involved in the process we are examining in a way not distracting to the viewers.

Finally, we present illustrations dedicated to a wider audience via television, games, and books where the data are "masked" by a more captivating, dynamic look (remember that they should still be based on correct assumptions even if not so easily spotted as in the previous representations). Every media is suitable for these representations, ranging from journal covers to augmented reality. My favorite examples are the wonderful videos of the WEHI Institute that gather hundreds of thousands of views on their YouTube channel. As already reported, this latter category is the focus of this book.

The types of illustration presented have the natural tendency to merge into each other. We cannot draw fixed boundaries between

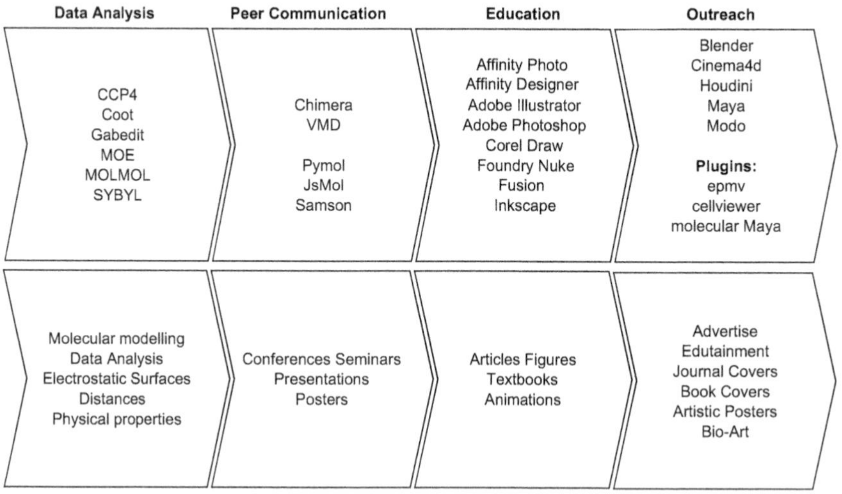

Data Analysis	Peer Communication	Education	Outreach
CCP4 Coot Gabedit MOE MOLMOL SYBYL	Chimera VMD Pymol JsMol Samson	Affinity Photo Affinity Designer Adobe Illustrator Adobe Photoshop Corel Draw Foundry Nuke Fusion Inkscape	Blender Cinema4d Houdini Maya Modo **Plugins:** epmv cellviewer molecular Maya
Molecular modelling Data Analysis Electrostatic Surfaces Distances Physical properties	Conferences Seminars Presentations Posters	Articles Figures Textbooks Animations	Advertise Edutainment Journal Covers Book Covers Artistic Posters Bio-Art

FIGURE 3.1 Visualization tools for a specific audience.

This figure is based on Figure 2 of Johnson and Hertig. 2014. "A Guide to the Visual Analysis and Communication of Biomolecular Structural Data." *Nature Reviews. Molecular Cell Biology* 15 (10). Nature Publishing Group: 690–98. doi:10.1038/nrm3874.

them, but it's still convenient to separate them in order to better select the tools we will use while building them.

Chemical language makes wide use of metaphors, as the purpose of molecular visualization is to support our understanding of the material world, making molecular structures and their interactions more intelligible.

Generally, atoms will represent the smallest part of our models, and we will use the model proposed by Dalton (in 1803!). A sphere will be more than enough for the level of detail and the scale of the objects we are going to represent even, of course, the way we depict the inner structure of atoms.

The radius of the sphere is proportional to the real dimensions of the elements. Atoms have dimensions smaller than the wavelength of visible light and thus have no color; however, for centuries, we have associated certain colors with elements based on their physicochemical properties.

One of the most-used conventions is the Corey Pauling Koltum (CPK), introduced by Corey and Pauling in 1953 and revised by Kolutm in 1965. In Table 3.1, you can find the colors coded from the original article.

We do not always need to differentiate the elements in a molecule, and often we rely on a different convention where we depict with the same colors only clusters of atoms in order to highlight an important site or property.

Today, thanks to the advance of CG, the use of a sphere to mimic the macroscopic material it constitutes is quite common, including metallic, plastic, and ceramic surfaces. I suggest caution and parsimony when employing this kind of visualization to avoid confusing your audience work.

Ball and Stick models made their first appearance in the 19th century; Kekulé was the first to introduce and use them. They were able to depict all of the information needed by inorganic chemistry and crystallography. In the '50s (almost one hundred years later), Pauling needed a model that could represent large and messy organic molecules and more accurately represent structure resolution; he introduced and developed a different type of chemical model called the "space filling model."

TABLE 3.1 Colors of the elements in the CPK convention

	Element	Color
	Hydrogen	White
	Oxygen	Red
	Nitrogen	Blue
	Carbon	Black
	Sulphur	Yellow
	Phosphorous	Pale yellow
	Fluorine	Pale green
	Chlorine	Green
	Bromine	Brown

(*Continued*)

TABLE 3.1 (Cont.)

	Element	Color
	Iodine	Violet
	Metal (both covalent and ionic)	Silver

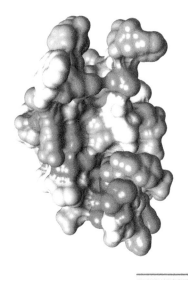

FIGURE 3.2 SIV protease crystallized with peptide product pdb:1yti.

The colors represent the hydrophobic surface of the molecule; the cylinder represents a scale bar of 10 Ångströms. The image was create using the software Chimera.

We know that when elements are in suitable conditions to react (depending on the physical properties of the element *per se* and on environmental condition such as pressure, temperatures, distance, etc.) they bond together to create different molecules. The energy needed for them to react is represented in different ways. Glowing or colored halos seem to be most often used to represent this concept.

It is well known that strict rules govern *which* atoms you can *bond* together and *how* you can bond them together. The software we use to draw our molecules generally checks how they can bond, so we do not need to worry about remembering those rules ourselves.

Generally, we represent the bond among atoms with wires or cylinders. Their colors and numbers depend on the nature of the bond.

The most common depiction for a *metallic bond* is realized by packing together spheres that represent atoms and connecting them

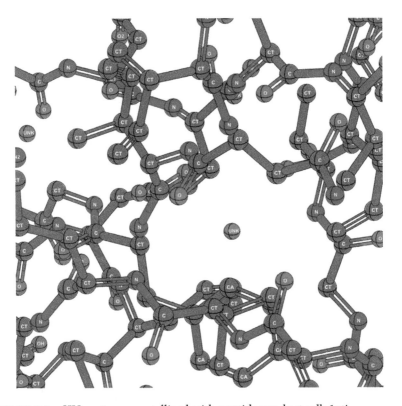

FIGURE 3.3   SIV protease crystallized with peptide product pdb:1yti.

Double and triple bonds are present in the molecule as seen in this image. This was created using Gabedit.

with cylinders with a height shorter than their radius. Emphasis is put on the uniformity of the metallic structure and on its packing.

When representing the *ionic bond*, we can use the same metaphor seen in the metallic bond and rely on the use of cylinders or wires. When we need to highlight a peculiar structure, we can simplify our representation using polyhedrons in which the vertices are the atoms and the edges represent the bonds shifting from an atomistic model to an abstract one.

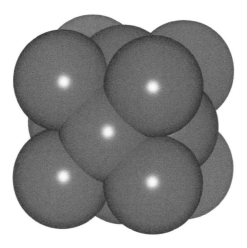

FIGURE 3.4 Representation of the standard orientation copper crystal shape made using Vesta.

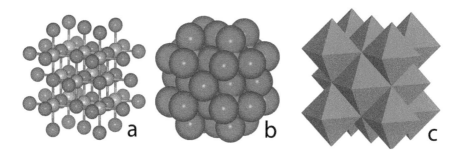

FIGURE 3.5 Examples of different representations of the structure of an NaCl molecule. (a) ball and stick, (b) space fill, (c) polygon style.

Carbon has a special place in chemistry since it is the element of life (although carbon can also be the constituent of inorganic substances). Due to it special nature, it has been given a dedicated branch of chemistry, organic chemistry.

Covalent bonds are ubiquitous in organic chemistry. They can be single, double, or triple bonds. Their depiction follows the same rules as inorganic bonds for what we consider small molecules, but their use is not practical for bigger biomolecules including protein, lipids, DNA and RNA, enzymes, etc.

Organic molecules are seen in a wide range of visualization methods that vary according to the intended analysis task. As already seen for inorganic structures, we can use atomistic or abstract models.

Apart from representing every atom (as a sphere), we can focus on bond-centric models as the *lines, licorice,* and *stick* models in which we depict only bonds using cylinders. We can augment this model using a *ball and stick* representation (as in the old fashion "real models" made of plastic). Instead of connecting atoms using cylinders we can improve

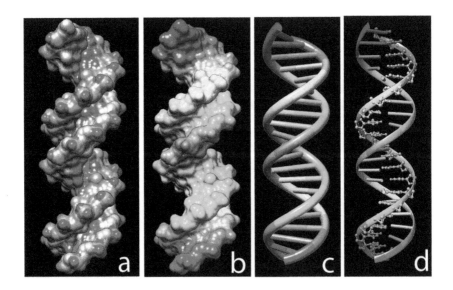

FIGURE 3.6 (a) wires representation, (b) ball and stick, (c) Van der Waals representation, (d) Specific DNA representation created to highlight base structure. All models were made in Chimera.

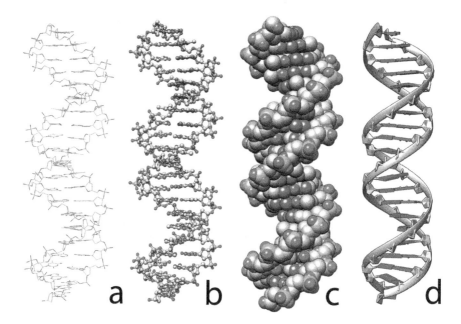

FIGURE 3.7 (a) surface representation. The colors are related to the hydrophobic properties, (b) Surface representation. Colors do not represent
any intrinsic properties of the molecule and are used to highlight its
morphology in an artistic way, (c) strands representation of the
molecule. It highlights the typical DNA ladder. Each color refers to
a base, (d) mixed representations. One strand is made of an
atomistic ball and stick representation; for the other a flat ribbon
representation is used.
All models are made in Chimera.

the representation using *hyperboloids* in the *Hyperballs* representation
introduced by Chaven.

Among abstract models, we can rely on space filling models (SFM),
Van Der Waals surfaces (VDW), solvent accessible surface (SAS),
solvent excluded surface (SES), molecule skin surface (MSS), ligand
excluded surfaces (LES), and convolution surface models (CS).

SFM and VDW models are simpler and most used since they are
built by merging all of the spheres representing the atoms of the
molecule. They depict the spatial volume that the molecules occupy.

SAS is the representation of all of the regions that can be accessed by a *solvent molecule* and is computed considering the solvent molecule as a single sphere. This metaphor is helpful for analyzing channels and binding partners. Its appearance is very similar to the VDW surface.

SES is a variation of SAS. Richards introduced it in 1977 and defined it as the union of all possible probe spheres that do not penetrate any atom in the molecule.

MSS is a variation of SES and makes possible the visualization of molecules with a few thousand atoms.

LES is a generalization of SES that instead of approximating the ligand with a sphere uses the geometry defined by VDW surfaces. It is more computational in comparison with SES and so far is used for static visualization.

CSS is the "scientific" name for *metaballs* or *blobby surfaces* in which we blend the potential of the atoms to represent an electronic density function.

Again, even when we deal with the representation of surfaces, *color* is an essential attribute.

Color can help us

- compare residues in the molecule and consequently the possible similar reactivity of the sites

- represent clusters with same physical-chemical property (e.g., a constant of dissociation in an acid environment such as pK_a, pK_b, charge polarity, hydrophobicity, etc.)

- highlight chains (the two chains that build DNA, a DNA/RNA hybrid ribbon. etc.)

While illustrating biomolecules, we will make use of other tools that can also help in the representation of their *architecture*. It is established that molecular architecture represents important features better than an atomist representation does. Richardson introduced the *cartoon* representation depicting the secondary structure as ribbons and arrows; it is still included in mainstream software.

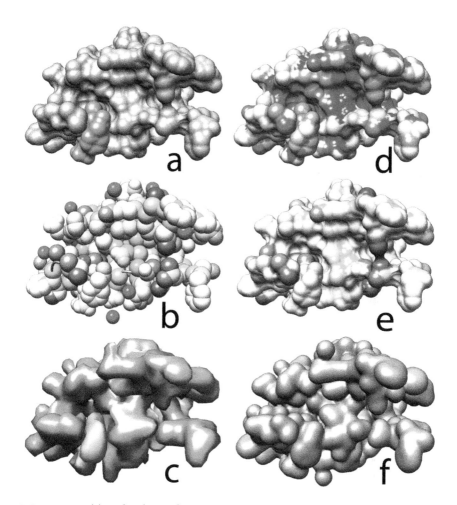

FIGURE 3.8   (a) molecular surface representation, (b) VDW representation, (c) LSS representation, (d) SAS surface representation. The solvent used for the calculation is water with a probe radius 1.4 Ångström, (e) SES surface representation. In this case, colors are used to highlight different properties in comparison with SAS, (f) Gaussian Surface, a representation belonging to the Convolution Surface Models family.

Simplification of surfaces is a technique used for a large molecular complex consisting of millions of atoms like a virus.

When we break the barrier between atomistic and mesoscopic scales, we need more tools. Due to the progress in physicochemical techniques, we

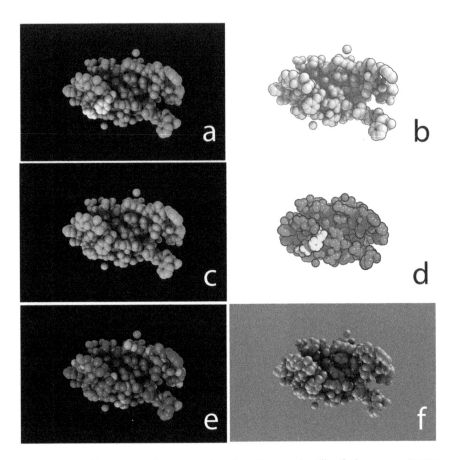

FIGURE 3.9   Representation created using Qutemol. All of these are VDW surfaces that use different colors to highlight features of the molecules (a) chains using two different colors, (b) monochrome representation to highlight the shape of the molecules, (c) "realistic" representation with occlusion surface shows the surface accessible to a solvent, (d) cartoon representation helps the viewer highlight the shape of the molecule, (e) VDW representation; every color refers to CPK convention, (f) a fake electronic microscope for an artistic look.

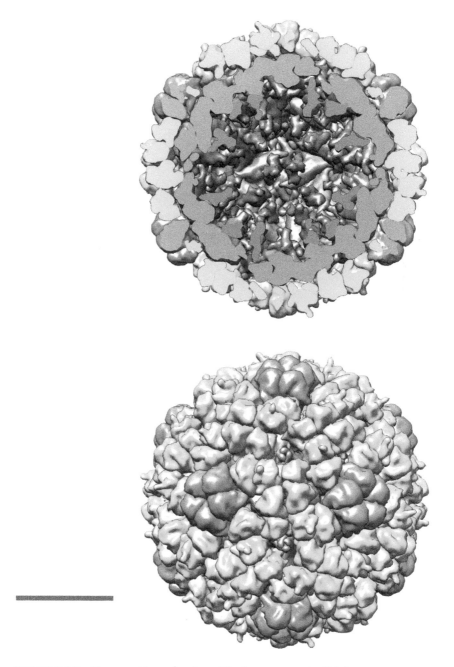

FIGURE 3.10   Cross-section of a virus. The bar represents 100 Angstroms.

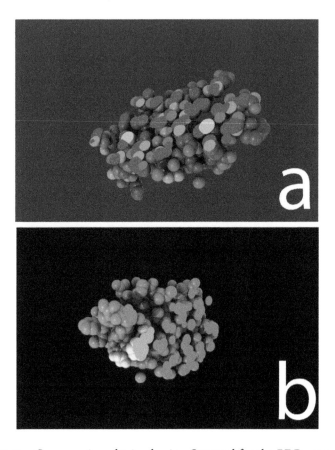

FIGURE 3.11   Cross-section obtained using Qutemol for the PDB entry 1yti.

are continuously narrowing the gap that separates biology and chemistry, and we start acquiring data in the range from nanometers to millimeters.

One paradigm is to keep simplifying the surfaces of the entities based on the magnification required to create models that can included several levels of detail. Another paradigm is to use a cross-section metaphor in a way analogous to cellular cross-sections.

So far we have presented the main metaphors that are (and should be) the starting points for creating more artistic interpretations merged with the visions from photography, cinema, and other media in order to narrate stories, the favorite human mode of learning.

The need of speculative pictures is still high due the old unanswered question: What is the living context in which nanomachines operate, and how do they operate in that contest? I am sure we will find an answer for these questions, and it will come from someone using creativity and imagination. Metaphors and tools for explaining scientific concepts are not written in stone, and we need them to evolve in order to make science progress. Let me conclude the chapter saying that even if sometimes we feel lost among the vastness of tools that emerge from the science and entertainment industries, we have to think of them as a great chance for freeing our minds and imagination. Even if they look too complicated at first, we need to remember that some of the best work starts from paper and pencil. I remember that the first pages in one of my first art history texts (too much time has passed; I don't have reference for that) described how the first hominids started drawing in their caverns with char, and that it was one of the first steps for in the growth of our imaginations. These early people also wanted to propitiate hunting since depicting an animal meant a better understanding of the prey. We have changed preys. Now they are viruses and bacteria, cells, etc. We are still playing with a charred stick and sometimes with other tools.

Questions

- Who is our audience, and what tools do we need to use in order to convey our message?

- What are the main metaphors for representing atoms and molecules?

- Are there strict rules for representing colors of the atoms of a molecule? What are the conventions used?

As an exercise, try to spot where and how the tools described in this chapter are used in any form of biomedical visualization in peer-reviewed journals. Would have you used other tools? Do they succeed in conveying their message?

SELECTED REFERENCES

Bonneau, Georges-Pierre, Thomas Ertl, and Gregory M Nielson. 2006. *Scientific Visualization: The Visual Extraction of Knowledge from Data*. Springer New York. Vol. 1. doi:10.1007/3-540-30790-7.

Briscoe, Mary Helen. 1990. *A Researcher's Guide to Scientific and Medical Illustrations*. Brock/Springer Series in Contemporary Bioscience.

de Ridder-Vignone, Kathryn D., and Michael Lynch. 2012. "Images and Imaginations: An Exploration of Nanotechnology Image Galleries." *Leonardo* 45 (5): 447–54.

Goodsell, David S. 2004. "Bionanotechnology: Lessons from Nature." *Nature* 430, 20. doi:10.1038/430020a.

Goodsell, David S. 2005. "Visual Methods from Atoms to Cells." *Structure* 13 (3): 347–54.

Goodsell, David S. 2009. "Escherichia Coli." *Biochemistry and Molecular Biology Education* 37 (6): 325–32. doi:10.1002/bmb.20345.

Goodsell, David S. 2011. "Miniseries: Illustrating the Machinery of Life: Eukaryotic Cell Panorama." *Biochemistry and Molecular Biology Education* 39 (2): 91–101. doi:10.1002/bmb.20494.

Goodsell, David S. 2012. "Illustrating the Machinery of Life: Viruses." *Biochemistry and Molecular Biology Education* 40 (5): 291–96. doi:10.1002/bmb.20636.

Jones, L. L., and R. M. Kelly. 2015. "Visualization: The Key to Understanding Chemistry Concepts." In *ACS Symposium Series* (Vol. 1208, pp. 121–140). http://doi.org/10.1021/bk-2015-1208.ch008

Keszei, Ernö. 2012. *Chemical Thermodynamics: An Introduction*. Springer, 2012. doi:10.1007/978-3-642-19864-9.

Kozlíková, B., M. Krone, M. Falk, et al. 2016. "Visualization of Biomolecular Structures: State of the Art Revisited." *Computer Graphics Forum* 36 (8): 178–204. doi:10.1111/cgf.13072.

Schubert, Thomas, Frank Friedmann, and Holger Regenbrecht. 1999. *Visual Representations and Interpretations*. doi:10.1007/978-1-4471-0563-3.

Wright, H. 2007. *Introduction to Scientific Visualization*. Springer.

Wu, Hsin Kai, and Priti Shah. 2004. "Exploring Visuospatial Thinking in Chemistry Learning." *Science Education* 88 (3): 465–492. doi:10.1002/sce.10126.

CHAPTER 4

Modeling and Lighting

T he first thing I do while lighting a scene is to turn off the default lighting in the scene. You don't have to rush while lighting; just add one light at a time in order to focus on the effect that each light has in the scene. We will start lighting a few of the models we created in the previous chapter.

Before starting to set up our scene, it is useful to understand how file formats for representing molecules are structured, in order to highlight their similarities with 3D object file formats. We will also introduce the software we will use for gathering the data we need to export to our 3D mainstream software.

As we have seen, scientific illustrators can rely on data collected at the atomic to mesoscopic level without losing continuity of information, thanks to several instrumental techniques of investigation. It will be our task to choose the best metaphors and tools to convey the information to our viewers and, when it is not possible to access any (or complete) data for our illustration, we will rely on drawing/building/creating our own representations.

We represent atoms as spheres since we do not need (generally) to cope with their electronic structure. Eventually, we can depict them as transparent clouds, but we will not calculate the shape of orbitals or the shapes of the bonds of the molecules involved in our illustration.

Molecules and macromolecules are represented by cylinders and spheres and by surfaces that look like metaballs (this term refers to the primitive shapes found in 3D software menus even if, in the previous chapter, I presented the different type of calculations performed to get coordinates of different surfaces).

When the size of the object of our depiction grows, we need to blur the surfaces that represent its building bricks; thus, we will use meta-balls, booleans, and volume representations and eventually create, modify, and sculpt the primitives present in our software.

ATOMS AND MOLECULES

As reported in a very useful text, *Chemoinformatics: a Textbook*, by Gasteiger and Engel (see suggested readings at the end of the chapter) two methods are commonly used for representing chemical structures in 3D space. One method uses an internal coordinate system describing the spatial arrangement of the atoms relative to each other. In this case, the internal coordinates refer to the bond length, bond angles, and torsion angles that can be arranged in the so-called Z-matrix. The other method, which is more intuitive for anyone versed in 3D graphics, relies on a Cartesian coordinate system, coding the x, y, and z coordinates of each atom.

Each approach is tailored for different tasks and calculations. The most common file formats describing the configuration, stereochemistry, and conformity use Cartesian coordinates (See Table 4.1).

In order to better understand how 3D coordinates work, we open a pdb file with a text editor. This is the pdb file for methane:

TABLE 4.1 Adapted from Gasteiger and Engel, *Chemoinformatics: a Textbook*, common file formats for representing structures.

File format	constitutions	configuration	3D structural/conformation
MDL	Yes	Yes	Yes
SMILES	Yes	Yes	Yes
SYBYL MOL2	Yes	Not explicitly	Yes
PDB	Yes	Not explicitly	Yes
XYZ	No	No	Yes

```
COMPND    METHANE
AUTHOR    GENERATED BY OPEN BABEL 2.3.90
HETATM    1 C  UNL    1    -7.164  2.898  0.000  1.00 0.00         C
HETATM    2 H  UNL    1    -6.094  2.898  0.000  1.00 0.00         H
HETATM    3 H  UNL    1    -7.521  2.448 -0.903  1.00 0.00         H
HETATM    4 H  UNL    1    -7.521  3.905  0.061  1.00 0.00         H
HETATM    5 H  UNL    1    -7.521  2.342  0.841  1.00 0.00         H
CONECT    1  2  3   4   5
CONECT    2  1
CONECT    3  1
CONECT    4  1
CONECT    5  1
MASTER      0    0    0    0    0    0    0    0    5    0    5    0
END
```

For a simple molecule (the easiest organic molecule we can write) like methane, the file is quite self-explanatory. In the first row, we have the name of the compound; the second row describes the author (in this case the software used for creating the molecule), the following 5 rows define the nature of the atom and its coordinates, and then we have the connections among atoms and finally the summary information.

Generally, we can download the 3D structure of the molecule we need to depict and do not need to create the file by ourselves from scratch; if needed, several types of software can help in this task.

Some of the mainstream 3D software has parsers for pdb files (The Foundry Modo® and SideFX Houdini® just to name two of them). This means that you can directly import pdb files and in a few seconds have a look at them in your viewport. Note that they do not generally offer the same importing options their stand-alone/plugin counterparts do. The most significant advantage is that the atoms will be depicted using the native primitives of the software used for parsing the file and take care of instancing and cloning our objects. A third method has several advantages: Use specialized plugins for your software to import your molecules. EPMV and molecular Maya are two examples. I've used EPMV in several works; probably the strongest feature is that it can create very light models letting you depict an incredible number of atoms. My advice on this topic is to explore the different choices and find the best path for you.

When I need to prepare a detailed illustration I organize my workflow without using plugins. I start focusing on the scientific part using a stand-alone scientific software (Avogadro, Chimera, Vesta and Nanoengineer) and then import the results on a mainstream 3D software.

When I need a quick schematic representation of my subject that require more time to be spend on the artistic aspect, I rely on 3D software that includes parsers (e.g., Modo or Houdini) or on embedded plugins inside the chosen 3D mainstream software. Plugins and parsers have the specific advantage of relying on internal modifiers and can make use of instancing and cloning.

I do not have a preferred workflow since every workflow can achieve the same results for a chosen representation. Plugins and parsers are invaluable, quick resources for creating content, but I urge you to get acquainted with dedicated scientific software; even if at a first glance you feel lost in all of their options, you will not regret spending time learning how to use them.

HOW WE DRAW VIRUSES AND NANOMACHINERIES

It is not easy to state a precise definition of nanomachinery; in this book, we use this definition:

1. *Inorganic macromolecules* that can act as small motors and actuators

2. *Organic macromolecules* that are involved in biochemical pathways

The first definition can include nanotubes and graphene sheets. The second definition is an intentionally very wide one and includes viruses, enzymes, DNA, RNA, co-factors, etc.

What software should we use?

INORGANIC MOLECULES

Avogadro is my first choice for drawing small molecules (Hanwell et al. in the suggested reference section). For depicting inorganic crystalline cells, Vesta is one of the best types of software you can find. It can read and save all formats currently used in physical chemistry/crystallography and if needed let you write your own molecule from scratch.

For special nanomachinery (mainly inorganic) assemblies, when the atoms that form molecules are in the thousands range, I use a specific dedicated software called nanoengineering. It is quite old, but its source code is available on github, ensuring the possibility of keeping its maintenance going. One of its strongest points is that, among all of the formats in the export option, the pdb file format is present. I suggest giving a look to SAMSON, a very promising software developed by NANO-D and Inria.

ORGANIC MOLECULES

Chimera and VMD are my choices when I need to draw organic molecules. Both have options for searching and fetching files from the protein data bank, where all files are stored in pdb format. Describing molecules by means of a pdb file is not unique; while fetching data from the protein data bank site you will notice that for some macromolecules (e.g., viruses), you have the option of downloading a pdb file obtained using electron microscopy. The file will probably have a link to the site European Molecular Biology Laboratory, "The home for big data in biology" (EMBL-EBI) and will be heavier to download than other pdb versions (Mb compared to Kb). Why?

This version of pdb files is obtained with imaging instead of crystallography techniques. The paradigm is quite simple. You take "pictures" of your subject at different height (Z-dimension), and you will finally have a stack of images (Z-stack) that let you recreate a surface. ImageJ, ICY, Fiji, and Chimera or VMD can handle Z-stack end export them in various format in order to let 3D mainstream software to import them as surfaces, volumes, or point clouds. We will be able to manipulate them in the same way we manipulate "traditional" 3D objects.

DRAWING YOUR FIRST MOLECULES

Small Organic Molecules for a Ball and Stick Representation
Open Avogadro (Figure 4.1)

a) Build>Fragment

b) Insert>Peptide>Ala

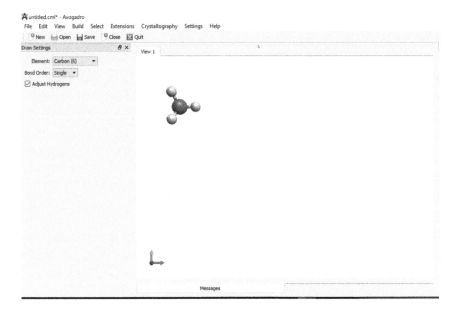

FIGURE 4.1   Graphic user interface of Avogadro.

c) Extension>Optimize Geometry (Ctrl+Alt+O)

d) File>Save as>Ala.pdb

We can check the file we created with a text editor.

```
COMPND  Ala
AUTHOR  GENERATED BY OPEN BABEL 2.3.90
ATOM      1 N    ALA A 1     0.045 -0.244 -0.066 1.00 0.00          N
ATOM      2 CA   ALA A 1     1.504 -0.041  0.011 1.00 0.00          C
ATOM      3 C    ALA A 1     1.951  0.046  1.482 1.00 0.00          C
ATOM      4 O    ALA A 1     1.196  0.271  2.419 1.00 0.00          O
ATOM      5 HN   ALA A 1    -0.374  0.135  0.791 1.00 0.00          H
ATOM      6 HA   ALA A 1     1.979 -0.918 -0.442 1.00 0.00          H
ATOM      7 CB   ALA A 1     1.916  1.226 -0.728 1.00 0.00          C
ATOM      8 1HB  ALA A 1     1.583  1.204 -1.771 1.00 0.00          H
ATOM      9 2HB  ALA A 1     1.499  2.121 -0.252 1.00 0.00          H
ATOM     10 3HB  ALA A 1     3.006  1.337 -0.732 1.00 0.00          H
ATOM     11 HN   ALA A 1    -0.356  0.292 -0.833 1.00 0.00          H
ATOM     12 OXT  ALA A 1     3.285 -0.112  1.653 1.00 0.00          O
ATOM     13 HO   ALA A 1     3.772 -0.297  0.829 1.00 0.00          H
CONECT    1 11       2    5
```

```
CONECT   2   1       3   6   7
CONECT   3   2       4  12
CONECT   4   3
CONECT   5   1
CONECT   6   2
CONECT   7   2       8   9  10
CONECT   8   7
CONECT   9   7
CONECT  10   7
CONECT  11   1
CONECT  12   3      13
CONECT  13  12
MASTER       0   0   0   0   0   0   0   0  13   0  13   0
END
```

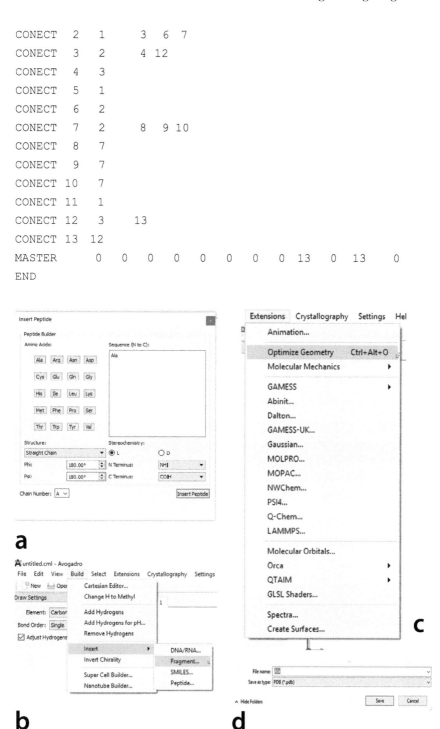

FIGURE 4.2   Step-by-step screen capture of the process described.

We can see that the pdb file reports all positions of the atoms and their connections. You can try to edit them to see how changes affect the representation when we later import it.

SIMPLE INORGANIC STRUCTURE

We start VESTA, then we

1) **File>New Structure**

A Dialog will open up. In this dialog window, we can provide the positioning and orientation of the layer. Also, the other tabs in this dialog let us insert the crystallographic parameter of our structure as *unit cell, structure parameter, volumetric data*, and *crystal shape*.

a) We will import one of the example structures; we can go to the menu **Import**

b), c) We choose from the **Vesta-Examples** directory **Al2O3.cif**, which is the crystallographic information file (cif) for Aluminum oxide.

FIGURE 4.3   Graphic user interface of Vesta.

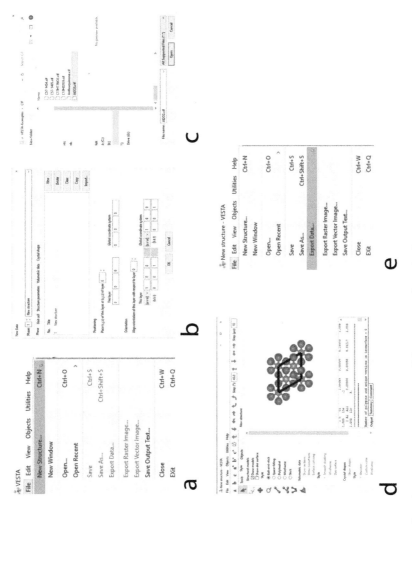

FIGURE 4.4 Step-by-step screen capture of the process described.

d), e) We will see the representation on the workspace and now go to **File>Export** data. This time we will use two file formats. One is pdb and the other is wrlm.

CARBON NANOTUBE

After opening the nanoengineer, we go to

a) **Tools>Build structure>Nanotubes**
Then a pencil will appear in the workspace.

b) We **drag the pencil** to create a nanotube of the desired length then click on the green tick on the upper right corner of the workspace.
Now we go to

c) **File>Menu>Export>Protein Data Bank**
And save the file as nanotube.pdb
Nanoengineer embeds a custom version of Qutemole for visualizing the molecules. It can be found in

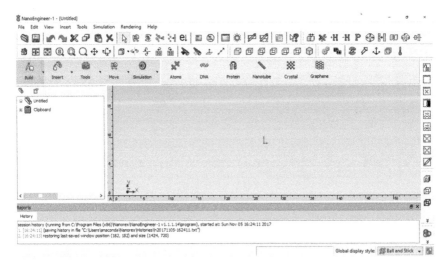

FIGURE 4.5   Graphic user interface of nanoengineer.

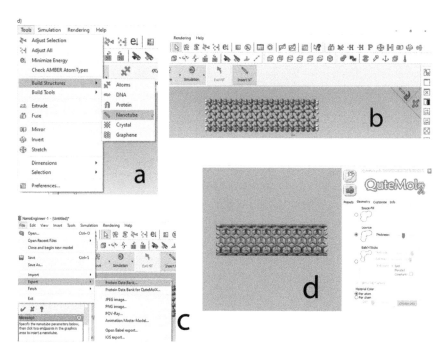

FIGURE 4.6   Step-by-step screen capture of the process described.

d) **Menu>Export Protein Data Bank** for QutemoleX. After clicking on the menu, we will see a new menu where there is written Launch QutemoleX.If we click on the button, QutemoleX will open visualizing the molecule we just created.

VIRUS

We will import the same virus using two different representations, one obtained via crystallography and one via microscopy (Cryo-Electronic Microscopy).

Fire up Chimera. (At the moment of writing, Chimera X, a more advanced version of Chimera, is in development. For this example and from now on, I will use Chimera since it is the most stable Chimera version available).

Importing pdb files in Chimera (coordinates obtained with crystallography)

First of all, we need to choose the virus we want to render. We will choose the Zika virus. We search for its entry. The PDB entry is **5IRE.**
We go to

a) **File>Fetch by ID**. In the dialogue window that will open, we write the PDB (biounit) entry (no matter if the caps are locked, PDB entries are not case sensitive)

Our virus will be visualized in the workspace (it can take a while depending on the dimension of the virus you choose to download).

This time we cannot use a ball and stick representation since it would look too overcrowded and would not highlight the feature we want to render later, so we will need to tweak the representation before exporting it.We go to

b) **Select, Select all** and then to

c) **Tools>High Order Structure>Multiscale Model>Make Models.**

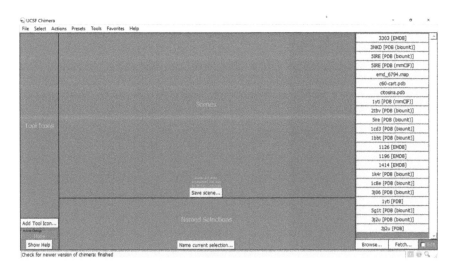

FIGURE 4.7   Graphic user interface of Chimera.

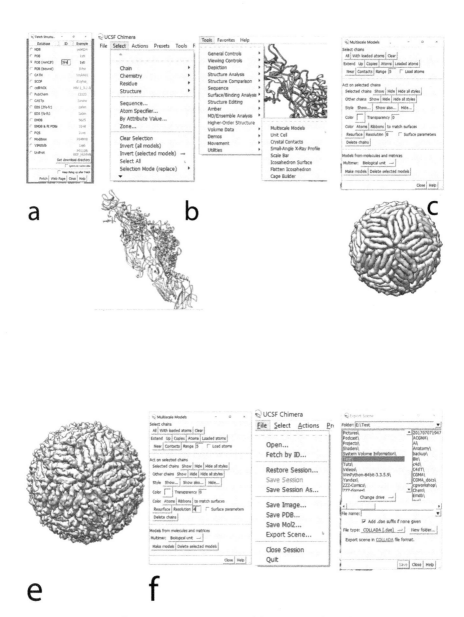

FIGURE 4.8   Step-by-step screen capture of the process described.

d) **Favorites>Model Panel:** We deselect the first entry and leave only the surface we just created. We go again to the **Multiscale Models,** and we **Resurface** changing the resolution to **4.** Now we can export the model with the following commands:

e) and f) **File->Export Scene** we select the DAE file format in order to keep as much information as possible. Note: I did not use the surface representation that can be found under **Action>Surface>Show** since it is not suggested for use with so many surfaces to visualize. If you have to represent a small molecule, you can try to split the model in its constituent chains and create a surface representation for each of them. If this fails, you can rely on the multiscale approach in which we just reported.

IMPORTING A PBD FROM EMBL-EBI (COORDINATES FROM CRYO-ELECTRONIC MICROSCOPY)

In order to import the correct file, we go to the EBI.ac.uk site. We can use two entries. One is EMDB 6794, and the other is the Fitted PDB structure 5y04.

a) **File>Fetch By ID** and then we insert our ID.

Note: EMDB maps can be quite big in comparison with conventional pdb files, so it can be good to download the files in our HD as well.

After opening our file of choice, the

b) **Volume viewer** will open. It works similarly to the levels command in 2D-imaging software. We can move the slider to **select the portion of volume** we would like to represent. When we are satisfied with the selection, we can save the results to the .dae format:

c) **File->Export Scene** we save our file as zika.dae.

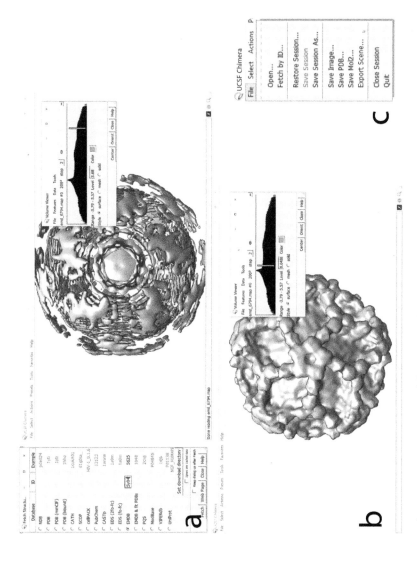

FIGURE 4.9 Step-by-step screen capture of the process described.

IMPORTING MOLECULES USING EPMV

After installing EPMV, following the step-by-step guide reported in the website, we will find a new menu in our 3D software of choice. After launching it, we will be presented with dialogue windows where we can choose to load a local pdb file or to fetch a pdb file from the protein databank. We choose to

a) **Import local pdb**

b) We use the alanine **file Ala.pdb** already created.

EPMV will take care of importing the alanine. Now we need to click on the representation of choice, **ball and stick.** We set up the representation, moving the sliders in the window that will pop up. While adjusting the settings of the molecule, our viewport will be updated in real time. This time there is no need to export our molecule and then import it in our 3D software, but it is wise to frequently our scene in order to avoid losing our job due to unwanted crashes.

Before Integrating Our Models in a 3D Scene

This section may seem redundant, but it is fundamental to understanding the key concept of 3D modeling. Even if you are already skilled in this field or you know that your model will be obtained by experimental techniques and so you will not need to model anything from scratch, I strongly suggest you read it.

Why bother with a chapter about polygon modeling (also NURBS (Non-Uniform Rational B-Splines) modeling) before importing our data and setting up our first 3D virtual studio?

The reason is simple: Even if we are dealing with data obtained with several physicochemical analytical techniques, our 3D software, in order to work with them, will "read" them as points in space that we can manipulate as polygons, NURBS, and (eventually) volumes. This is true for every level of abstraction. In other words, it is true for everything from the first methane molecule we created with spheres and cylinders to a DNA/enzyme assembly, where DNA chains are represented as 3D "arrows" and the enzyme as a metaball.

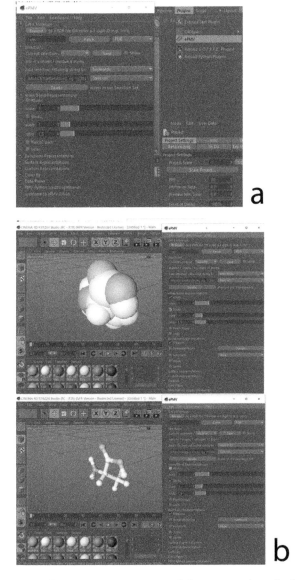

FIGURE 4.10   Step-by-step screen capture of the process described.

One of my favorite approaches to learning modeling is Andrew Paquette's. The first chapter of his book, *Computer Graphics for Artist*, contains a very important reminder quoted from Stemberg in 2003:

> *CG professionals are aware that 3D applications change on a nearly annual basis and every few years the most popular application will have changed to something new. When this happens, artists who understand computer graphics for what it is will be in a much better position than those who look at it from the more limited perspective of the applications they are familiar with. Applications are relatively easy to learn in comparison to the expertise that comes from practice and understanding.*

We will follow the same approach, introducing general concepts common to all 3D applications.

CREATE A 3D COORDINATES SYSTEM

The first step inside a 3D space is to set up a coordinate system. The natural choice is to rely on a Cartesian space. Our software will take care of it. Software generally represents a 3D Cartesian space and

FIGURE 4.11 Cinema4D Interface.

FIGURE 4.12   Modo Interface.

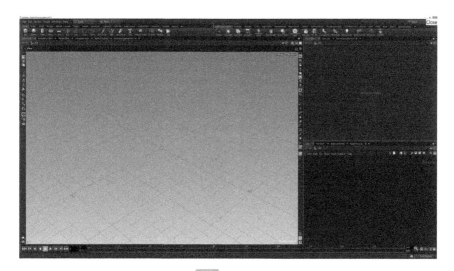

FIGURE 4.13   Houdini Interface.

the corresponding planes with grids. In the origin (the point where all the plane intersect, that corresponds to the center of our viewport) of 3D space we will find three colored arrows. They are red, green, and blue arrows representing x, y, and z directions, generally arranged according to the right-hand rule. We should check the convention used by the software since this is not always true.

Just a refresher: Hold your right hand in such a way that your fingers curl from the positive x direction to the positive y direction through the 90° angle, and your thumb will point to the z-axis. Once the 3D origin

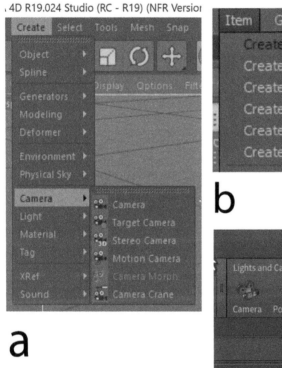
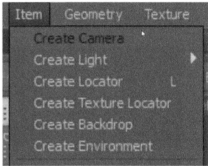

FIGURE 4.14 How to create a camera a. Cinema4D, b. Modo, c. Houdini.

FIGURE 4.15 Effect of using different Focal lengths: 16, 24, 35, 50, 80, 135, and 200 mm. The image used is a render test project that can be found in the Foundry Modo software examples directory.

FIGURE 4.15 (Continued).

is established and before starting to draw, we need to learn how to navigate in this space. Every 3D software presents a top view and side view; front and perspective views are by default in preset layouts.

If we want to further customize the viewport, we can create a virtual camera (**menu-> create** or **menu->item**) and tune its settings. The most common are

- Focal length
- Projection (perspective, orthographic, isometric)

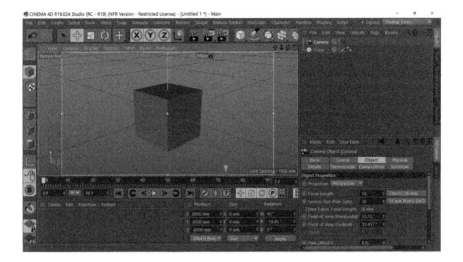

FIGURE 4.16   Effect of using projection: orthographic and perspective camera; notice how the lines on the plane are affected by the chosen camera.

- Sensor size

- Field of views (horizontal and vertical)

- Dimension of the film gate

- Focus distance

FIGURE 4.17 Effect depth of field (DOF): 0.02, 0.04, 0.08, 0.1, 0.2, 0.5, and 1. It is important to notice how DOF can help us in grabbing the attention of the viewer.

FIGURE 4.17 (Continued).

- F stop
- Shutter speed

Changing these settings affects how the software will render the final image/movie, and a preview is commonly available in the viewport. We can select it as our main view of choice and then move, dolly, or pan in the same way as with a real camera.

FIGURE 4.18   Effect of Film Gate: CCD 1", CCD 1/2", 1/3", 2/3", 4/3" DSLR, BlackBox Camera, Black Magic 4K and full 35 mm 4K.

Notice that experimenting with different film gates can improve the aspect of our final image. While working in a laboratory not all the sensors are default "classic" 35 mm.

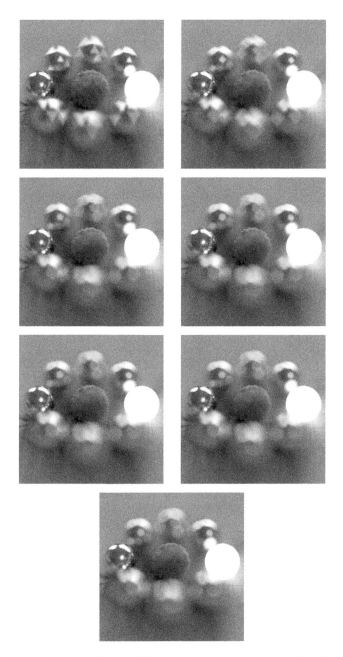

FIGURE 4.19   Number of Iris Blade: 3, 4, 5, 6, 7, 8, and circle. The effect is only visible while shooting using a very wide aperture. 🔊

CREATING 3D OBJECT(S)

The second step is to create an object in our 3D world.

There are several ways to do this. We have already created the coordinates of the atoms/molecules we will represent using specific software; we can go to the import menu and

- create it vertex by vertex,

- import it from another application or repository,

- import or draw splines that will guide the creation of our objects using specific modifiers, or

- insert a "primitive" with the push of a button and then modify it.

Primitives can be geometric or readymade simple objects that can be tweaked using special menus present in the software and then used as the initial building block of a scene. Before inserting the model into the 3D space, we need to define the parts of the model called elements or components.

The 3D elements of a model are

1. Vertex

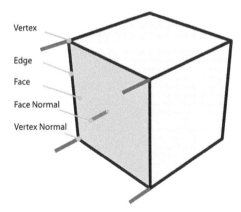

FIGURE 4.20   A primitive with 3D elements highlighted.

2. Edges

3. Faces

4. UVs

5. Normals

Vertex(es) or vertices represent points in space. They have 3D coordinates but no dimensions, and we cannot render them. When we connect two points, we describe a line. Connecting *two vertex pairs* of the same object creates an *edge*. Edges cannot connect more than one point at time. This is very important, since if something goes wrong while importing models, the model will be corrupted. While connecting all of the vertices in the correct order, we will define the *faces* of the object (mesh) that we want to represent. They cannot be rendered, since they have no dimensions. The order in which the vertexes are connected can also be visualized in the viewport.

Faces occupy a space in 3D and so they can be rendered. We also need to know their orientation in order to make light bounce off correctly; so we will add a vector to each face. *Normals* are generally drawn starting from the center of each face. Also, in order to create smooth transitions between faces that share edges, we generally calculate average normals in each edge.

After creating our object, we will generally change its color and other attributes (sticking an image to it, for example). To do this we need another set of coordinates called UV coordinates for each face of our object. This new set of coordinates can be edited in a texture editor in order to check and change the visibility of an image attached to the object.

BOX MODELING

In our application, we look for a menu for creating primitives

Create>primitives

and create a cube. Contextually a menu will open where we can enter all parameters for our cube. The representation in the 3D viewport will vary according to our software. All 3D software has a wireframe mode, a constant shading or flat mode, a quick render mode, and a more detailed render mode.

FIGURE 4.21 Create a primitive in Cinema4D, Modo, and Houdini a. Cinema4D, b. Modo, c. Houdini.

- Wireframe represent an object in line only. It draws vertices and then connects them with lines. We can generally choose to see all of the lines of your mesh in a sort of x-ray mode or to hide lines according to their relation to the camera.

- Constant shading is a representation in which all of the faces look flat.

- Quick shading is a fast representation with lights and textures and a more accurate representation that can include almost all of the features of the final render (shaders, textures, depth of field, etc.)

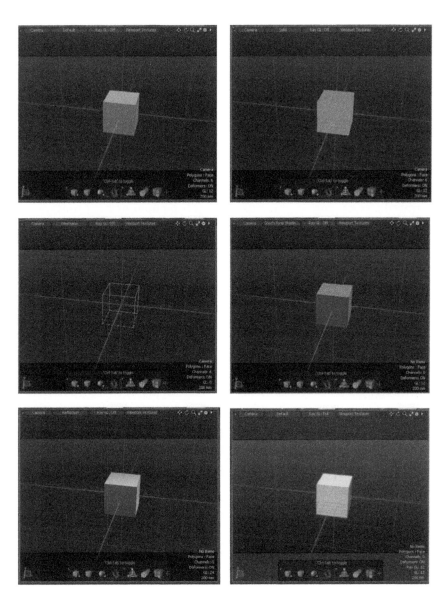

FIGURE 4.22 Different viewport settings in Modo. Starting from the top left: default, flat, wireframe, gooch, reflection and advanced GL full.

Each of these representations will be useful at different stages of the modeling process, even if every artist has a personal favorite mode of viewing. My personal choice is a quick shading with wireframe superimposed and the back facing edge hidden.

Software nowadays also features interactive preview rendering that creates a fast rendering draft of your scene able to be changed and tweaked in real time. We have already seen that software represents points in a 3D space. We have a 3D matrix, so, even without knowing calculus, we can *translate* and *rotate* it in space applying a *transformation* matrix to our mesh. We can apply a transformation matrix only if we have a *reference*. We need a pivot point (represented generally by three arrows), and we can change its position to obtain transformations depending on its position.

Generally, the default location of the transformation matrix, the pivot point, is the center of the object. When we animate an object, a *state matrix* is stored for each frame of the animation, and it is used to modify the position, size, and rotation value of the object over time.

Let us see what kind of tool we can use if we want to modify the primitive we created.

We can *select* the vertices, edge, or faces using the same tools we find in a 2D paint application (square, circle, lazo, and freehand selection, just to name a few).

The selected element generally changes color, and a *gizmo* (similar to the pivot) will appear. At this point, we can move, rotate, resize, and translate it according to its properties.

While *moving, rotating, resizing* and *translating* elements we can rely on snapping (generally the icon snap is a variation of a magnet) our movement according to

- Grid line intersections
- Curves
- Single vertex
- Center of edge
- Center of face
- Center of object

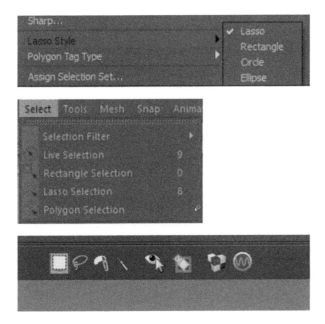

FIGURE 4.23 Selection options in Cinema4D, Modo, and Houdini a. Cinema4D, b. Modo, c. Houdini.

Working from primitives guarantees the artist's ability to make models that are always solid structures, without gaps caused by missing faces or faces that are not properly joined. On the negative side, box modeling will tend to create more faces than necessary in comparison with other ways of modeling. For our purposes, it will be a workhorse tool while creating simple representations of nanomachineries.

DIRECT INPUT

Another way of modeling is by *direct input*.

We define the exterior boundary of an object by clicking the position of each of its points. This way of modeling is limited to flat objects and used only in creating flat surfaces that we can then extrude and bend in the case of more complex object.

Part modeling is a close variation of box modeling. Each part of the object is modeled separately and then merged together. Widely used, it is preferred to box modeling for very complex models. After creating/ importing the first polygons in the scene, we will rely on an extensive toolkit of modifiers in order to pursue our vision.

Modification of Faces of the Mesh

- *Add or delete*
  The first operation we can perform on the polygons created is to add/delete vertices, edges, and faces.

- *Subdivision*
  There are several ways to subdivide an object. Every kind of software will include a *catmull-clark* subdivision methodology and other proprietary solutions. Subdivision is used in order to obtain *more* polygons to move; it will also be useful for smoothing your mesh. Comparing 3D sculpting with sculpting with clay, we can say that *subdivide* is the process of smoothing the corners of our draft. In the same way, we can choose how to round angles.

- *Extrusion*
  Extrusion detaches the selected face or faces from its/their neighboring faces and then connects the two groups with new faces. It works like adding clay to a sculpture in a selected part of a model (in this case a face) in order to have more material for modeling a bump in the surface of the object.

- *Cutting*
  We can cut our model with a virtual knife, creating new edges and vertices. Several ways of cutting are general allowed, and the software takes care of creating additional vertices, faces, and edges.
  Our mesh can also be

- *twisted*

- *bent*

- *exploded/fractured* or

- *melted*

Every modifier will have a specific range of influences that can be scaled and moved according to our need. Two special operations you can perform on your mesh are *decimation* and *remeshing*.

Decimation will take care of reducing the polygon count of our model, while r*emeshing*, depending on the software used, will eventually reduce the polygon count, trying to keep an optimal mesh geometry according to the number and positions of the faces.

Modification of the Edges

- *Bevel* shaves the selected edges with a straight curve or smooth curves along the selected curves.

We can modify our meshes with the help of spline. Spline is based on the original spline, long strips of wood used by shipbuilders to draw smooth curves. They were adjusted by the rotation of pegs on the floor of large design lofts. 3D software lets us create polygonal objects without being affected by the resolution and lets us align vertices along aligned curves. Generally, it is used when working in orthographic views. The big advantage of using spline is that we can define multiple polygons at once with smooth results and even distribution of *vertices*.

Working with modifiers and curves lets us create several kinds of surfaces. The most common modifiers are

- *lofting* (creating a polymesh starting from two meshes)

- *extruding* (connecting two curves with polygons)

- *Revolving* (similar to extrude, but with an end result like working with a lathe) In some software, the modifier is called *lather*. *Face filling* (fills the curves with a single face)

Vertex modifiers
We will see which commands can be applied to vertex.

- *Adding and deleting*
  It is important to notice that some software will delete the selected vertex or vertices without affecting surrounding vertices or edges.

- *Collapsing*
  Deletes the vertices and fills the gap with a vertex in the average position of the adjacent ones

- *Welding*
  Unlike other commands, this has several names from one application to another. It takes care of merging together vertices. It is possible to retain the position of one of the two vertices or to create the new merged one on the line that connect the two.

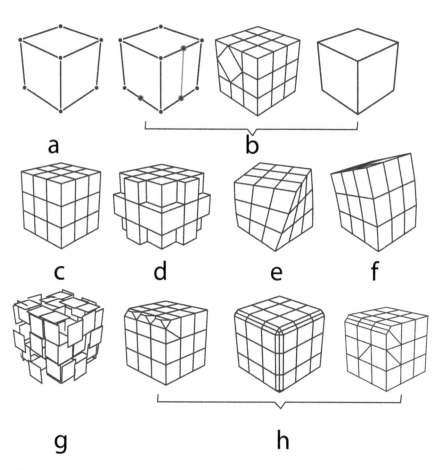

FIGURE 4.24 a. Cube, b. add and delete vertex, edge or face, c. subdivide, d. extrude, e. twist, f. bend, g. explode, h. blevel vertex, three edges and single edge.

Modifiers on whole meshes

Two or more meshes can be combined using Boolean options thus subtracting, intersecting, and merging them. Software has improved the results obtained from Boolean operations, even if in some industries

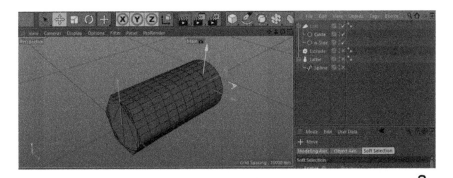

a

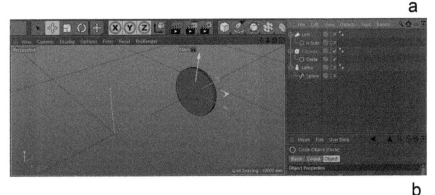

b

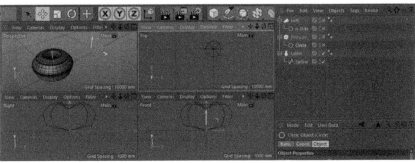

c

FIGURE 4.25   Cinema4D Modifiers a. Loft, b. Extrude, c. Lathe.

a

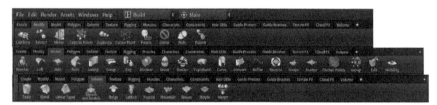

b

FIGURE 4.26 Menu related to the modeling actions presented in a. Cinema4D, b. Modo, c. Houdini.

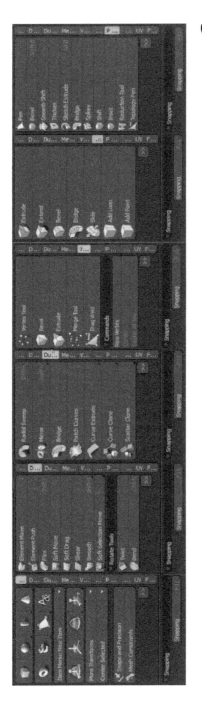

FIGURE 4.26 (Continued).

(games as an example) they are generally avoided for good reason. The first is that they can create polygons with more than four vertices, *n-gons* that do not work well when subdivided. We will try to use them sparingly. In Figure 4.24 we can see a summary of the operations presented.

NURBS

What are NURBS? I remember that the first time I used 3D software I really didn't understand why there were two different kind of menus for creating the same geometry. There was a NURBS menu and a polygonal menu, but when I clicked one of them I just saw a sphere in the center of my viewport.

Of course, that was not true; there is a difference. Maya has two menus with icons that looks very similar except for their color. We will not use NURBS a lot in our work, but it is useful to know what

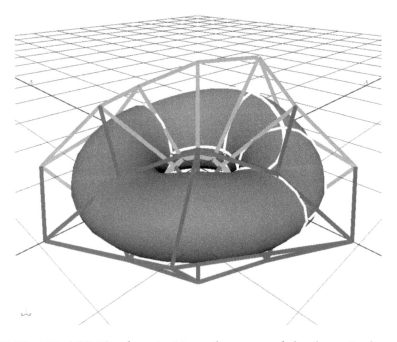

FIGURE 4.27 NURBS sphere in Maya, the aspect of the donut is changed moving the spline control points.

they are. NURBS is the acronym for **non-uniform r**ational **B**ezier spline. We have already described splines. *Bezier* refers to Pierre Bezier, a French automobile designer who first published his results in a paper in 1986; *rational* refers to the fact that the curve can be rational depending on weighted values at each control point. These are *non-uniform* since they are divided based on known positions. NURBS curves come in a variety of degrees.

A *first degree* is a straight line; a *second degree* has a third control point and the degrees increase as control points are added. A *third degree* spline is generally used to make a tangent with another curve or surface. The majority of manmade objects you were designed using five to seven degree curves in order to control accuracy.

In NURBS modeling we create the curves that work as guide for our software to build patches. When we need to modify a NURBS, we drag around the knots of our spline instead of each vertex (useful with the increasing complexity of the model). These tasks require careful planning since they have to satisfy a few requirements that are not needed in polygon modeling.

- Their direction must be correct.

- Their knot definitions must be correct.

- Endpoints of two curves MUST be coincident, and they must be tangent or there will be a visible tangent break between the surfaces.

If you are used to polygon modeling, you will probably find that some of the parameters in NURBS modeling are not editable, specifically

- Texture coordinates (we will describe later what textures are; for now just keep in mind that they behave "differently" for NURBS.)

- Normals (they can only be reversed)

Texture coordinates and normals (apart from reversal) are not editable because these properties are embedded in the surface.

What Are the Advantages of Modeling Using NURBS?

First of all, if we want to create an object to be manufactured we will be probably use CAD files.

- CAD software is NURBS based. NURBS are great when we need to import a CAD file with the description of an instrument or laboratory equipment to create an illustration.

- NURBS have more restrictions but are more precise for manufacturing.

- You can subdivide NURBS without thinking about subsequent problems due to the subdivision process.

- NURBS carry less polygon cost.

- We can create any kind of trimmed, cut surface, spline on surface without having to deal with the problems of Boolean or incorrect snapping.

These are a few of the applications and advantages of using NURBS that should kindle your curiosity to try them. I have discovered their versatility while making use of the Autodesk Fusion 360 when I had to recreate a layman representation of a thermal analysis instrument. I still use spline modeling when I need to work on precise mockups of instrument diagrams and when I need to create fast prototypes.

Paradigm of Procedural Modeling

A whole book will not be sufficient to describe how to work with procedural modeling. I have to say that I've started using procedural modeling quite recently and have just scratched the surface of it. *Why did I start learning it?* If you are working in science there is the chance that you know and use programming and metaprogramming languages such as Python, R, Matlab, and Labview. Software and environments that use scripting and procedural modeling have the ability to speed up your scientific workflow and tedious calculations on data. They can be invaluable for modeling, solving, and simulating complex *systems* (the term system here is used broadly) based on coded rules. First of all, a lot of mainstream 3D software includes a console that writes down all of the actions you are performing. If you have been using a computer for eons, you will probably remember working without a graphic user interface that makes use of windows. The 3D software that you use today merges all of the functions you need for creating your 3D world and all of your action. In other words, when you create a sphere in 3D view, you call commands. If you open a C4d Python script log after creating a sphere primitive you will find the following lines of code:

```
import c4d
from c4d import documents, plugins
#Welcome to the world of Python
def main():
    c4d.CallCommand(300000116, 300000116) # Script Log...
    c4d.CallCommand(5160, 5160) # Sphere
if __name__=='__main__':
    main()
    c4d.EventAdd()
```

The same is true for Maya and for almost any 3D software. This is important because you can do the same thing "in reverse." In writing this command, you can create the sphere primitive and modify any of its parameters. Every command you use via the GUI can be controlled via scripting. If, as we said before, you use a programming language in your work, you will be able to

- use your 3D software to plot your data

- animate the objects you created according to data acquired via a technique that lets you export the coordinates in a data matrix

- create particles according to strict rules and then animate them

- simulate a growing process using an emitter and creating a metablob of the particle created

- manipulate volume data before plotting them as a volume object

- tweak the aspects of every single model in your scene without going through each one (and of course for thousands and thousands of objects)

Although most software can be *scripted*, my choices are Houdini, Cinema4D Mograph, and the procedural modeling tools inside Modo. My suggestion is to learn SideFX Houdini even the learning curve can seem quite steep. Learn by using the scripts that are present in the software community. You will be astonished to see the range of applications in the scientific community, including the following:

- growing of plants

- differential growing models

- animations of plots based on econometric data

- the depiction of all the flight routes in a day in our planet

- recreating the surface of a planet using satellite data

- simulating the interactions between a gas and a liquid, etc.

Study Questions

Focusing on the cover of a scientific journal (*Science, Nature, American Scientist, New Scientist, Advanced Materials*, etc.)

- What kinds of metaphors are used to convey the ideas to the reader?

- Is the cover based on real data or on artists' abstract ideas?

- Would you have used the same tools used by the artist?

- Would you be able to recreate the models you see via box modeling?

- Which techniques do you think were used? Are they manufactured by extruding, revolving a surface, sculpting?

- How many objects are present in the scene? Do you think that it would be too time consuming to set up each of them?

- Do you think there are data available in public databases on the object of depiction?

SUGGESTED READING

Daniele, T. 2009. Poly-modeling with 3ds Max: Thinking Outside of the Box. Focal Press/Elsevier.

Gasteiger, J., and T. Engel. 2003. *Chemoinformatics*: a textbook. Wiley-VCH.

Goddard, Thomas D., Conrad C. Huang, and Thomas E. Ferrin. 2005. "Software Extensions to UCSF Chimera for Interactive Visualization of Large Molecular Assemblies." *Structure* 13 (3): 473–82. doi:10.1016/j.str.2005.01.006.

Hanwell, Marcus D., Donald E. Curtis, David C. Lonie, Tim Vandermeerschd, Eva Zurek, and Geoffrey R. Hutchison. 2012. "Avogadro: An Advanced Semantic Chemical Editor, Visualization, and Analysis Platform." *Journal of Cheminformatics* 4 (8): 1–17. doi:10.1186/1758-2946-4-17.

Paquette, A. 2013. *An Introduction to Computer Graphics for Artists*. Springer. https://doi.org/10.1007/978-1-4471-5100-5

Shirley, P., and S. Marschner. 2009. *Fundamentals of Computer Graphics, 3rd Ed.* https://doi.org/10.1177/004057368303900411.

CHAPTER 5

Scene Setup

NUMBER OF OBJECTS

Before seeing how to setup the lighting stage for the models we created, I would like to share a bit of advice. We saw in the introduction that we are free to make use of different levels of abstraction. New techniques let us gather enough information, from the atomic scale to the scale of cells and tissues, to depict very realistic representation of the nanostructure. Is it always worth it? It depends. One of the most impressive and (at least for me) puzzling *constants* in chemistry is Avogadro's number. Its definition states that one mole of a substance (the equivalent weight or the weight in grams corresponding to the atomic mass of an element) contains $6.022140857 \times 10^{23}$ units (atoms or molecules depending on the substance in exam). Thus, the only way to accurately represent anything at a nanoscale is crowded and not always practical. First we need to focus the viewer's attention on the subject of our illustration. An overcrowded illustration won't help us in this task. Second, even with the fastest computer, it is always better to keep our scene small for ease of rendering the image. Even if we are using a real time engine for our task, keeping a small memory footprint is necessary to keep the viewer's experience as smooth as possible.

SCALE OF OBJECTS REPRESENTED

Another very important point related to the number of objects to depict in our scene is the *scale* of the objects represented. If a scene is not scaled properly, a lot of issues can take place while modeling and rendering. The most common problems are

- wrong intensity of the light (while setting the light and rendering the final take)

- wrong rendering of the surfaces of the models due to wrong thickness

- irresponsive viewport or crash of viewport due to objects' being too big or too small

Also, generally simulations are scale based, and a different scale can drastically affect the behavior of the simulation. This is especially true while simulating fluids. Ripples or tendrils can look unrealistic and the simulation can become irresponsive during calculations. Also be careful to apply correctly scaled deformers to avoid erratic behaviors.

Finally, be sure to set a proper scale before dealing with particles simulations.

The first thing you should do before starting a project is to set the scale of your project and units. Every 3D software has this option, and 3D software generally deals with "real" units.

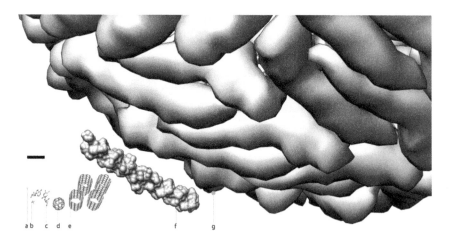

FIGURE 5.1   Starting from the left, depiction of (a) methane atom, (b) alanine, (c) fullerene C60, (d) carbon nanotube, (e) a collagen filaments, cryo-EM structure of Zika Virus (pdb:5ire) The scale bar represent $10 \text{ Å} = 10 \cdot 10^{-10}$ m.

Gather as many references as possible to learn the correct quantity and sizes of the objects to be illustrated. Apart from the usual sources (books and search engines such as google scholar JSTOR, scopus, etc.) a website I always consult is the BioNumbers database (bionumbers. hms.harvard.edu). Its About Us section states

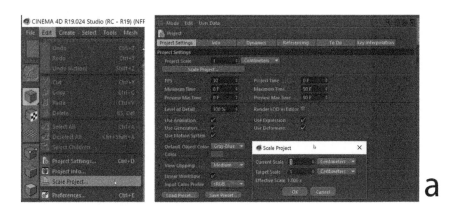

FIGURE 5.2 Setting up the scale of your scene in a) Cinema4D b) Modo c) Houdini.

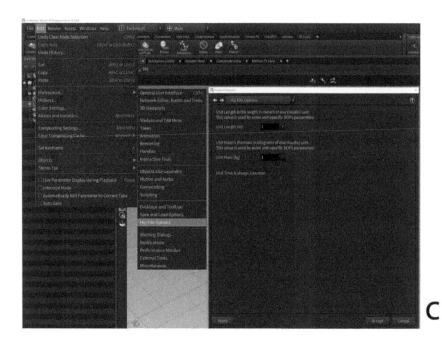

FIGURE 5.2 (Continued).

In BioNumbers we aim to enable you to find in one minute any useful molecular biology number that can be important for your research.

and

*It is our hope that the database will **facilitate quantitative analysis** and reasoning in a field of research where numbers tend to be "soft" and difficult to vouch for.*

The website contains a huge database of useful numbers, starting from how many cells are in a human body to how many white cells per cm^3 we can find in blood. I cannot stress enough that the first thing you need are very sound scientific roots for the concepts you are depicting.

SETTING UP LIGHTS

First, we get rid of all of the lights in our scene. Yes, you read correctly. Your 3D software probably already includes *default lights*; get rid of them and start to test one at a time the range of lights in your toolkits.

The scene we are going to use is very simple. Instead of using a sphere as a primitive object, we will use a reconstruction of a virus obtained by x-ray crystallography. The scene test is set up with one directional light and a Zika virus obj file (surface obtained using Chimera software after importing the pdb entry 5ire). We turn off all other lights and leave only a directional light with default settings.

Before tweaking the light settings, one word about shadows. Our software is capable of using different algorithms for calculations. As a rule of thumb, *raytraced shadows* are sharp, precise shadows and have a longer rendering time. *Mapped shadows* are tweakable soft shadows. The higher the resolution of the map, the higher the definition of your shadow and the quantity of memory you are going to use while rendering your image.

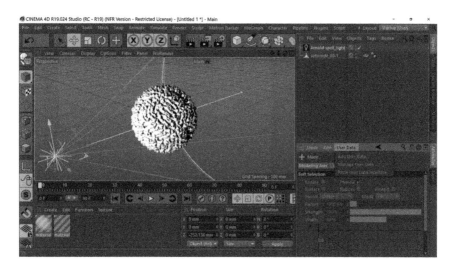

FIGURE 5.3   Wireframe of the scene.

We will use raytraced shadows to test the different types of light present in our software. This is my default choice while illustrating stills.

After opening the menu, we see lots of choices for light items. It is quite easy to get lost in the process, so I follow this very simple strategy:

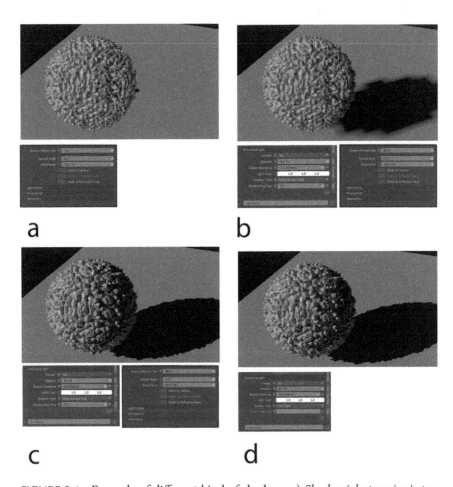

FIGURE 5.4 Example of different kind of shadows a) *Shadow/photon size* is too small. That is why we do not see any shadow. b) Now the dimension is ok, but we still have problem with the *shadow map res* (resolution) c) We increased the shadow maps. Things now are improving; keep in mind that the memory allocated for calculating the shadow increased. d) Raytraced shadows. They look sharper. In this case it is not necessary to tweak the *shadow map res*.

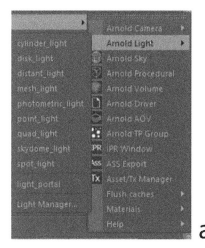

a

b

c

FIGURE 5.5   Light Menu. a) Arnold inside Cinema4D b) Modo c) Houdini.

1) If I need to create emphasis on the scientific aspect (i.e., on the clarity of the process/reactions), I try to make the setup as simple as possible. I use as few light sources as possible so fewer shadows will distract the viewer. A few area lights or a sky light or *global illumination* is sufficient for our representation. Remember that the focus is on the *shape* of what we are representing. Two examples are inorganic repetitive structure and diagrams that explain the process involved in building materials at an atomic scale.

2) When I need to create emphasis only on a portion of the scene to communicate the complexity of the environment depicted, I simulate a small studio where the model is my nanosubject. In this case, the tools are mainly spotlights, omni lights with low intensity to fill the dark spots and one or more backlights (kick lights) to enhance the silhouette of the subject. A more creative approach

is the norm here. I will use the tricks used in other fields of computer graphics: volumetric atmospheres great for creating an underwater look and are used for depicting the inside of the body. Spotlights in combination with volumetric lights create what is called *cathedral lights* or *God Lights* and a sense of a vivid and realistic micro world. Of course, we can complete in post-production all the details needed (sparks of lights, relighting, grading, whatever can do the job).

Each software is different and will offer a customized menu, but you can generally adjust the following parameters while setting up your light.

1) Kind of light (the source type)

2) Shadows (adjust the quality settings and choose color and transparency)

3) Light intensity (including kind of decay)

4) Light colors

5) Volumetric or not

Let us see them in detail and how their settings affect our renders:

2) Kind of Light

It can be seem obvious, but it is important to think about how light works in real life. One of the most common errors in illustration is to use a spotlight without keeping in mind that the shape of the shadow depends on this parameter.

Why does it matter? Shadows always give a hint of the surroundings of the focus of your scene. Even if they are not directly visible, they will change the overall appearance of your image, giving the viewer a hint of the dimension of the object and the way it is set up in your 3D space.

In order to give you a practical example, refer to Figure 5.6.

In the first scene, we have only our Zika virus. In the second, the Zika virus is beyond a matrix of small particles that are not seen by the

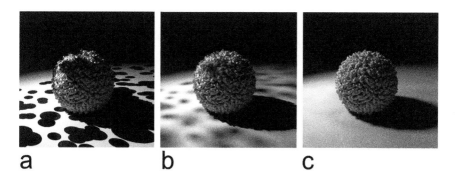

a   b   c

FIGURE 5.6   a) Example of sources of light the source of light is a point light with no dimension. The object near the source of light projects a visible shadow giving a hint of the environment of the scene. b) the dimension of the source of light is bigger than in a) and the shadows look more diffuse. c) the source of light is even larger; shadows are almost invisible but contribute to the overall atmosphere of the image.

camera but project their shadows on the virus, giving a hint of a more crowded environment. This would not be my choice for a simple clean scene, but it can add details in a more artistic rendition.

Shadows

We have already seen a few of the available settings for shadows. There are, of course, more for tweaking their aspects. I would like to focus further on the shadow transparency and eventual shadow colors. The first can be changed to make your scene appear less crowded. Fewer shadows, in the kind of scene we illustrate, mean more clarity. Also, in real life, you do not see pitch black shadows, and you should take care of that even changing this setting. Changing the color of your shadow can be helpful while rendering transparent and waxy material in order to cheat their aspects, making them more artistic. Shadow color can be helpful as a quick and dirty solution to light that is reflected from other objects on the shadow of your subject. For example, if you put an object in a red room and light it with a spot, in real life, the walls of the room (depending on the materials of the walls and the intensity of the light) will give a red nuance to the shadow of your object. You can cheat this effect by toning the shadow a bit.

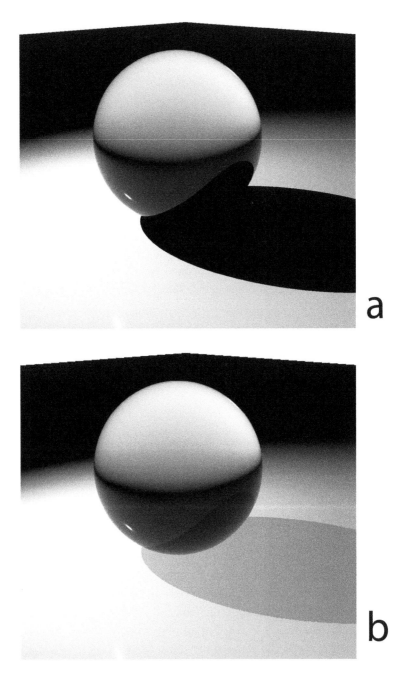

FIGURE 5.7 A fast neutral material applied to the surface of our object. Tweaking the transparency of the shadow gives the image a lighter appearance a) transparency off b) transparency on.

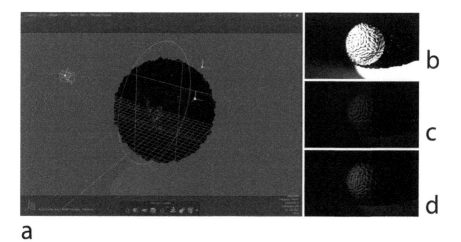

FIGURE 5.8   Example of different kinds of light decay a) Scene setup b) no decay c) inverse decay d) inverse square decay. In order to make the subject visible, the light intensity has been increased 1000x.

Light Intensity

Please note that light intensity in the real world is not linearly proportional to the intensity of the source (i.e., if you double the intensity of your source it will not have the same intensity at double the distance). Light decays in a way stronger than we usually think. 3D software keeps this in mind, and generally, the default setting is linear decay. If you want a more "realistic" approach, you can change your settings and restore the correct behavior (quadratic effect).

Light Colors

We know that changing the color of a scene can change its mood drastically. Think about cold and warm scenes, alienish green scenes, and ubercontrasted and dark thriller movies. Grading a movie is a full work, and achieving the final look is one of the most important tasks for making the audience understand its mood (or change of mood) even before it consciously understands it. Color of lights is a very extensive topic that can be the topic of a standalone book. I want to give practical advice here. Generally, you do not need to change it drastically. A red

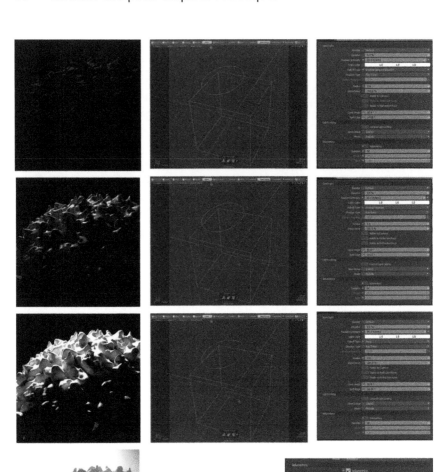

FIGURE 5.9 From the top of the image to the bottom. Lighting setup: Inverse Distance Squared, inverse distance, none, volume metrics on, three neutral light setup, red and blue three studio light setup, setting with rim light for subsurface scattering materials.

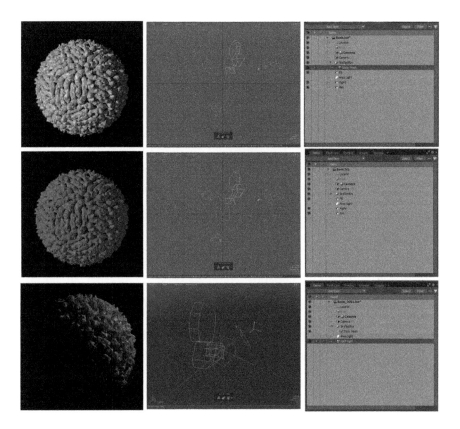

FIGURE 5.9 (Continued).

light is just slightly red. If you use pure red (i.e., rgb 255,0,0), it will appear quite strong in your overall image. I generally change the light settings to Kelvin temperature by steps of 100 K to test different takes.

Volumetric or Not

At this point, adding volumetric effects slows your renders, so keep this in mind before using volumetric lights. In the real world, lights always interact with our atmosphere; ideally, we always use volumetric rendering with just a bit of atmosphere. You use volumetric effect to simulate a very cloudy and foggy environment. That generally does not happen at a submicron scale, so why bother? The reason is that volumetric effects simulate the idea that your scene is "underwater" or better, inside a liquid environment. Actually in biology the ubiquitous solvent is water and so 100% of *invivo* chemical reaction takes place "in water."

Be sure to accurately check the scale of your project and the dimension of your source of light to avoid unwanted and unexpected results.

Since I generally prefer to teach by example, Fig. 5.9 et seq illustrate the same scene with different takes. As you can see, each kind of lighting creates its own mood, and a combination of all of the tools we have is necessary to achieve the desired result. Unfortunately, there is no "make it wonderful" setting on our software (at least not yet), but you can try to guide the viewer's attention and mood depending on the aim of your illustration.

COLORS

The Fig 5.9, 5.10 and 5.11, illustrate tools dedicated to color selection in Cinema4D software. Analogous color pickers are present in all 3D software. What are the general options for selecting colors?

a) Red, green, blue (RGB), hue, saturation, value (HSV) and Kelvin (K). These icons activate the related color sliders.

b) Icons are used for selecting complementary, analogous, split complementary, and tetrad colors and are equiangular in the color circle.

c) Circular color pickers and color slider selection

d) Sliders for the HSV

e) Kelvin and temperature slider color selectors

f) RGB slider selectors

g) Palette selection

In the previous section, we shared a few words about the effect of lighting on the mood of your overall scene. This is even truer (if possible) for colors. The choice of the color of your image is of capital importance for conveying the intended message to your viewers and in order to focus their attention. First, remember that we all have different perceptions of colors. If you are in a congress and are showing your

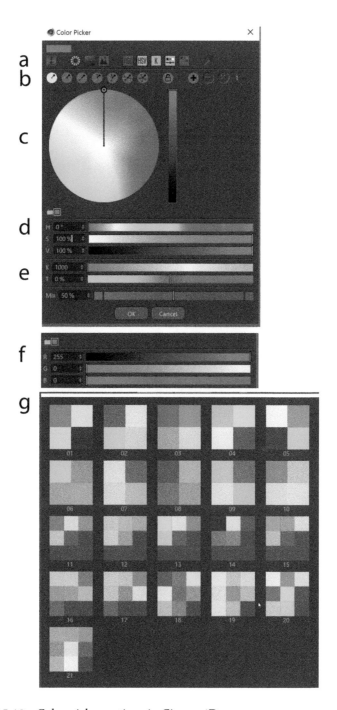

FIGURE 5.10   Color picker options in Cinema4D.

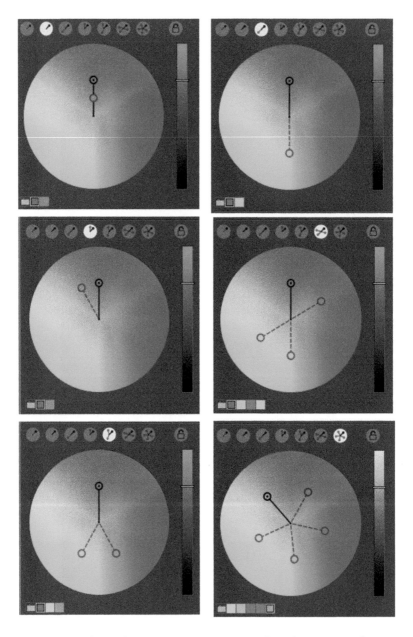

FIGURE 5.11   Color picker options in Cinema4D for selecting complementary, analogous, split complementary, tetrad colors and equiangular in the color circle.

audience a beautiful scheme about a novel route of synthesis, the safest choice is to use a colorblind safe palette.

I've learned the hard way by preparing what I thought were beautiful illustrations for a report and being told by my supervisor that he could not see the differences between two samples. Puzzled, I told him that they were quite different colors. He responded, "I see; for me they look the same, probably because I suffer from deuteranomaly."

The second factor is the impact a choice of color has on the mood of your audience. There is a wide selection of literature on the topic, and many different opinions on the topic. Finally, you must select the media for representing your rendering/illustration. To decide, answer the following questions:

1) Will your image be printed in a scientific journal or a book?

2) Are you creating an animation/movie?

3) Will stills from your animation be used for creating flyers?

4) Will your final media be a screen?

You are probably familiar with the recommendations for publishing figures in scientific journals. Everyone should understand what dots per inch of your image (dpi) means and why to use the highest resolution possible while sending your images to the publishers. While creating digital images we use *pointillism*. We can think of our image as composed of millions of dots. Our software helps us choose the correct value for that dot. More dots give a more defined image. If you would like your plots to look smooth and not jagged or pixelated in the final version of your work you should be sure to have rendered the most appropriate number of dots.

After the dots are computed, it is important to convey their information depending on the canvas. Sources that emit light will, of course, look brighter and more saturated and will need a lower resolution than printed sources. Printed works also subtract light to create color and are not additive source of light. Screens create images by mixing RGB dots. There are different ways to assemble these small sources of light, and during the years these sources of light have changed in nature, but the basic ideas have not changed.

FIGURE 5.12   The image on the right simulates deuteranopia (top), protanoto-
pia, tritanotopia (bottom).

After rendering our image, we can save it to a file format that records
the information for each pixel without taking any shortcuts (read
compressed file formats like gif, jpg). These files are called bitmap
format (tiff, tga, bmp, etc.).

When the media is printed, we use another way to create color. We
subtract color using layers of inks. Generally, the layers are cyan,
magenta, yellow and black. We still use dots, this time printed on
paper (or plastic, metal, etc.). The more information and dots we have,
the better in all cases presented. This is true for black and white images
and colored images. remember that, for example, when we need to print
black and white images we need more dpi than in colored images.

In 2D illustration, vectorial formats are *the* standard. Vectorial
formats (svg, png, ai, etc.) save the information of the created shapes
and colors as formulas. The huge advantage here is the ability to
recreate images as big as needed without losing quality (i.e., instead of

creating the new pixel for creating a circle, you save the formula for creating a circle and when you need to print it in a huge format, you have your computer create the needed pixel).

What about mobile phones, tablets, ebooks? Here as well, the higher the resolution the better, keeping in mind that mobiles and tablets have small screens with high resolution that can reach printed paper and also that e-ink devices are similar to printed paper.

How can we be sure that the color we see on the screen will be printed? In order to avoid problems, you should work with a calibrated monitor, render your images at the highest possible resolution, and, when possible, make use of vectorial file formats instead of bitmapped formats. Also, we need to calibrate every peripheral in order to adopt a calibrated workflow that starts from our monitor and ends with the printed paper (or final digital media). This subject could fill a book. However, considering that a printed image is at least 5 cm x 5 cm you should proceed in the following way:

a) Calibrate your monitor and printer regularly.

b) Render your image in HD (High Definition) or UHD (Ultra High Definition) resolution (minimum 1920 × 1200).

c) Save it using bmp, tga, or tiff format and try to avoid jpg. Vectorial formats work well, so if you can produce eps, svg, png it will be fine.

d) After your rendering is done, import your image into software that can change its color space from Red, Green, and Blue (RGB) to Cyan, Magenta, Yellow, and Black (CMYK) and check whether the image you obtain is close to the results you had hoped to obtain.

In the compositing section, we will briefly talk about how to composite your image. This process involves rendering each component of your image (the diffuse light, reflected light, the info about the space position of each object, the depth of field of camera, etc.).

In order to make it work correctly you should always use *linear workflow*. All of the settings presented apply to animation. If possible, fill your file with as much information as possible and compress only as the very last step of your final rendering.

Color and the Viewer

Now that we have tried to ensure that the color in our screen will be reasonably close to that of the final result, we need to think about how color influences the viewer and what kind of tricks we can use to obtain our desired results. The first tip is that eyes love bright colors; they will first look at the bright spots of your image and then to the dark corners. So keep your subject well lit.

1. Create contrast in your image, both light and color contrast. If you use contrasted colors, the viewer will immediately spot the differences among objects in your scene. If you use variations of the same colors, your image will look uniform (and with luck not flat).

2. Using strong primary colors creates sharp contrast and strong emotions. We have already seen that the use of contrast depends on the context (would you eat a blue tomato?), but in the biomedical field and "science" in general is true that

   a. Very dark/brown colors help represent diseases. A rule of thumb is to think about your nails. If they are total black (and you didn't use any nail polish) something is wrong with you! Also, zombie colors represent nasty nanomachines. One wouldn't color a tumor pink or pastel colors.

   b. Green colors can be fresh and natural; there are also *green chemistry* and *green molecules*, but also a green cell among pink cells (or a green molecule among pink molecules) can represent something to worry about. Again, a zombie green is related to illness.

   c. Red can be used for everything related to tissues and blood. Live tissues are represented as very red tissues, full of life. If you have looked at an anatomy book with real pictures, blood generally is a bit darker and tissues do not look saturated. Red also, since it is a primary color, will capture the attention of the viewer.

   d. Pastel colors, low contrast combinations, pink, azure, orange, etc., are quite neutral.

So how does one choose? I generally choose a photograph that makes me feel the emotion that I would like to convey to the viewers and create a palette with its colors. Finally, you should experiment with breaking all of the rules, and create your own style, trying to keep consistency among your illustrations while keeping scientific accuracy. As reported earlier, colors are not present in the real world at the scale of our project, so (if you are not using an established convention such as CPK), strictly speaking you are free to experiment without risks.

STUDY QUESTIONS

Focus on illustrations you can find in scientific journals.

- What differences are there in the lighting used for the cover and that for the graphical abstract?

- How many sources of light are used?

- Do you think that the kind of source light used in the cover and in the abstract is the same?

- What colors are used for the graphical abstract and what colors are used for the cover?

- Choose at least 50 photos you like (not only in the scientific field). What are the main sources of light in the images? What colors are used? Sample the colors you like in order to create a palette for each photo.

- Try to choose 10 eye-catching advertisements in scientific journals. What colors are used? Are they primary, complementary, secondary, etc.? What emotions do the different palettes used convey to the viewer?

SUGGESTED READING

Birn, J. 2014. *[Digital] Lighting and Rendering.* (K. Johnson, Ed.) Third Edition. New Riders.

Brinckmann, C., and K. Korpershoek. 2014. *Color and Empathy: Essays on Two Aspects of Film*. Fourth Edition. Amsterdam University Press.

de Leeuw, Ben. 1997. *Digital Cinematography*. AP Professional.

Foster, Jerod. 2014. *Color: A Photographer's Guide to Directing the Eye, Creating Visual Depth, and Conveying Emotion*. Peachpit Press.

Shimamura, Arthur P., and Stephen E. Palmer. 2011. *Aesthetic Science: Connecting Minds, Brains, and Experience*. Oxford University Press.

CHAPTER 6

Rendering

HOW LIGHT INTERACTS WITH MATERIALS AND WHY IT MATTERS

In my everyday work, I am lucky to have access to an electronic microscope. Readers probably already know that it can take pictures at thousands of orders of magnitude and get details of the surfaces of samples on the order of nanometers. It is very similar to an optical microscope, but the lens is electromagnetic, and instead of visible light that spans its wavelength in the field of 400–700 nm, we use higher frequencies and smaller wavelengths. This difference will probably not be noticed by the user since both optical and electronic microscopes interact with the user via monitors that show the sample being examined. Electronic microscopes are portable and do not need much space; a desktop is enough. I digress about electronic microscopes because I would like the reader to focus on the idea that we record the interaction of matter with light at several different wavelengths. This interaction will be responsible for the final aspect of the object we are going to study: If a wavelength is not perceived as colored, the true image acquired with an electronic microscope is black and white. We would artificially color it in order show the features of interest, but this is part of the necessary elaboration of the images taken.

What path do rays of light follow from a source of light (the sun) to the retina? After leaving the source of light, they travel quickly (8 minutes from the sun) to the surfaces of the objects around us before being recorded by our eyes. They interact with these surfaces in different ways. The material of the surfaces reflects, diffuses, and scatters the rays. The roughness of the surfaces influences how much and how the light interacts with the material. If we strictly follow the rules that we have extrapolated from the observations of natural phenomena, we need to keep in mind that the energy of our rays do not increase during this path and needs to be conserved in some way.

How do we simulate these processes? We need to describe the material we want to reproduce using different models that can take care of each of the possible interaction that materials show in real world. These aspects are merged in what is called a *shader*.

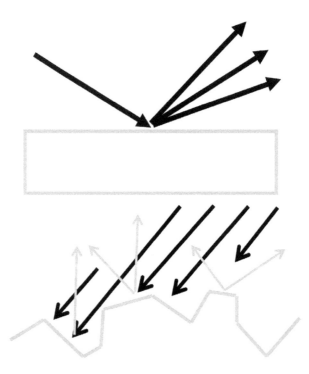

FIGURE 6.1 Reflection and diffusion of light on a smooth surface (top) and rough surface (bottom).

Notice the self-shadowing effect in the second case.

Software has different options for defining the aspect of a material, but they all adopt as a common background the *principled shaders* that were at first introduced by Pixar.

Pixar developed a physically based, art-directable shader centered on an intuitive approach with as few parameters as possible. Each parameter is designed to be in a range from zero to one (even if this can be pushed beyond its limits for artistic reasons), and all parameters are designed to be as robust and plausible as possible.

The parameters are the following

1) *baseColor* (surface color) the diffuse color of the surface

2) *subsurface* controls the diffuse effect using a subsurface approximation (behavior of milk, skin, wax, etc.)

3) *metallic* the metallicness of the surface. In this model, metals have no diffuse component but have a tinted incident specular

4) *specular* incident specular amount

5) *specularTint* to give artist control on the color of incident specular toward the base color

6) *roughness* controls specular and diffuse response roughness *sheen* grazing component that is intended for cloth. It is very useful when we need to imitate electronic microscope images

7) *sheenTint* amount of tint sheen towards base color

8) *clearcoat* special purpose specular layer simulates a varnish on your objects

9) *clearcoatGloss* controls clearcoat glossiness. 0 is related to a satin appearance while 1 corresponds to a gloss appearance.

As you can see, the principled shader frame covers almost any aspect of the interaction light matter giving enough freedom to the artist to experiment in the domain of non-physical behavior.

In order to explain how changing these parameters influences the aspect of our shader, we can see the examples in Figures 6.3 to 6.5.

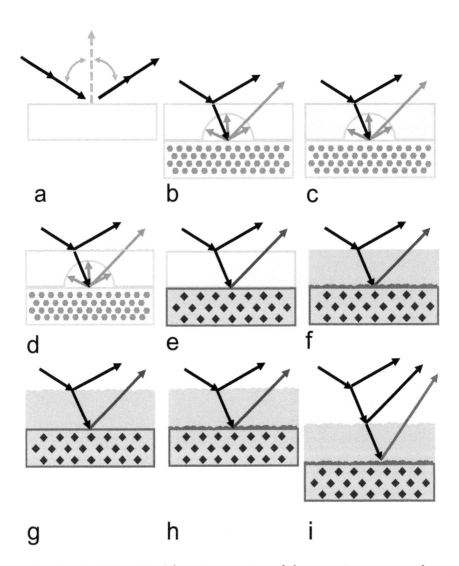

FIGURE 6.2A AND 6.2B   Schematic overview of the most important surface scattering as show by Weidlich and Wilkie in the seminal work "Arbitrarily layered micro-facet surfaces" in: *Proceedings of the 5th international conference on Computer graphics and interactive techniques in Australia and Southeast Asia.* ACM, 2007. pp. 171–178 and on the LM Pixar principled shader a) reflection on a smooth surface b) glossy paint c) tinted glazing, d) frosted paint, e) metal foil, f) metallic paint g) frosted metal h) patina i) multi-layer.

— Smooth
  Rough
☐ Clear
  Tinted
  Diffuse Uncolored
  Diffuse Colored
▦ Metal

FIGURE 6.2A AND 6.2B   (continued).

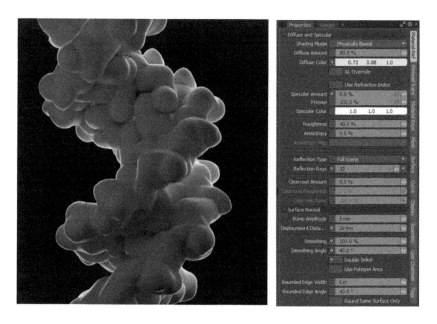

FIGURE 6.3   From top to bottom we increased the specular amount of the shader leaving all other parameters set as default. In the top image, the aspect of the object (a surface obtained from a DNA sequence) has a plastic aspect.

Increasing the reflectivity confers on the object a more metallic appearance.

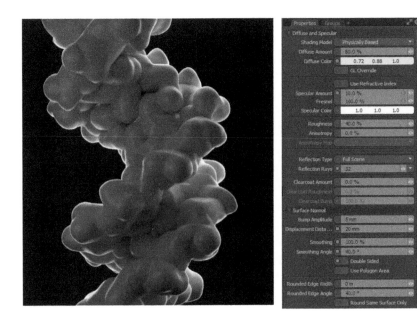

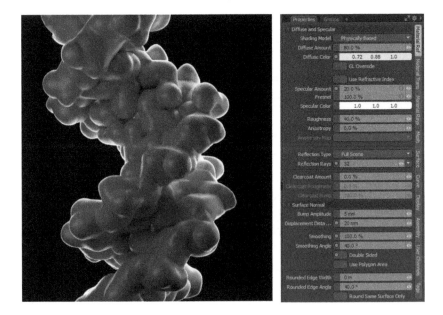

FIGURE 6.3 (continued).

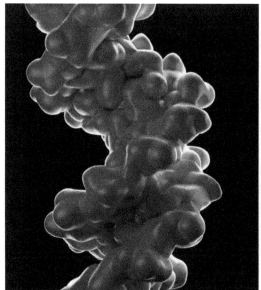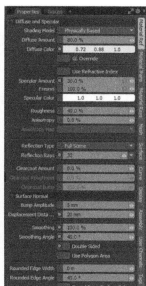

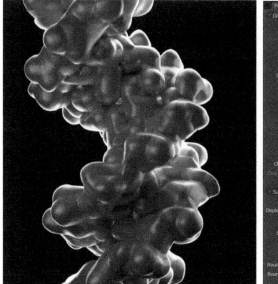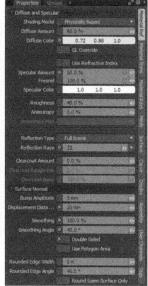

FIGURE 6.3 (continued).

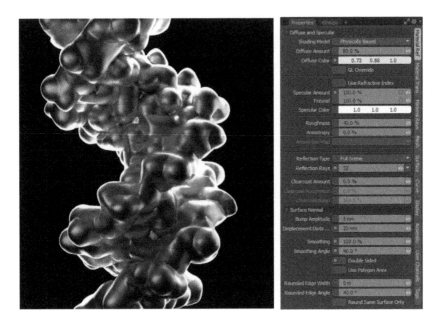

FIGURE 6.3   (continued).

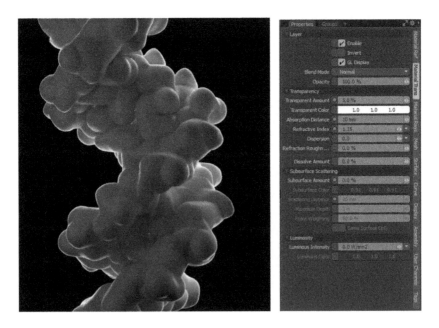

FIGURE 6.4   From top to bottom we increased the transparency amount of the shader leaving all other parameters set as default.

In the top image, the aspect of the object (a surface obtained from a DNA sequence) has a plastic aspect. Increasing the reflectivity confers on the object a more glassy appearance.

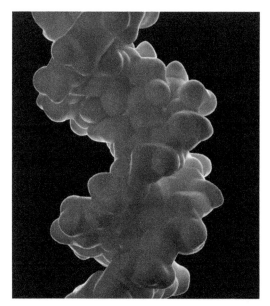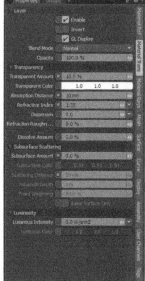

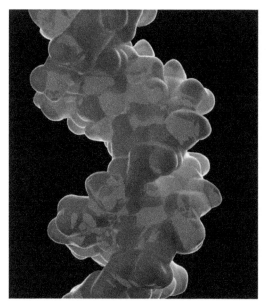

FIGURE 6.4 (continued).

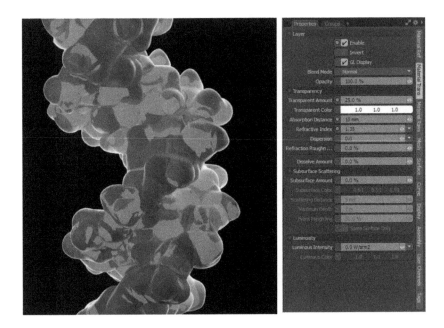

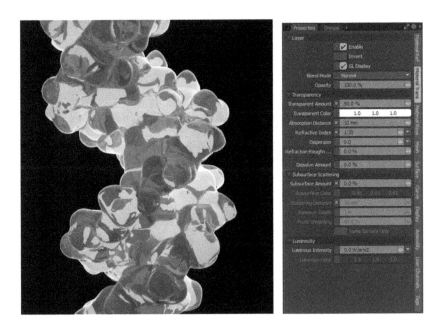

FIGURE 6.4 (continued).

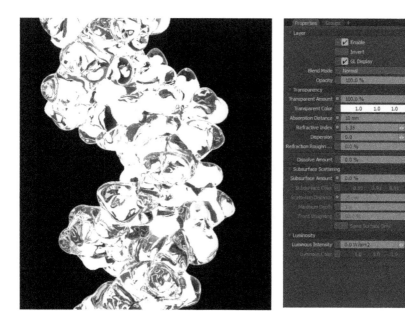

FIGURE 6.4   (continued).

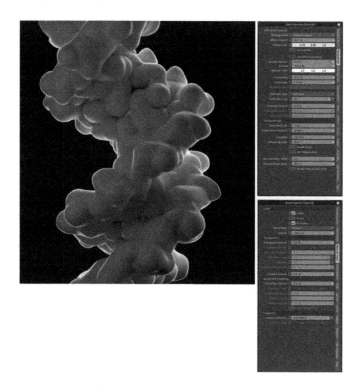

FIGURE 6.5   (continued).

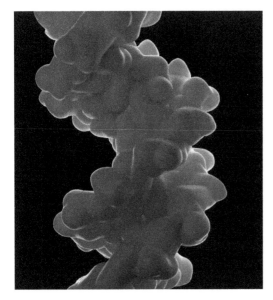
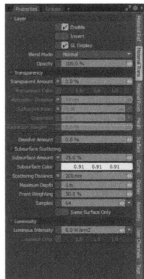

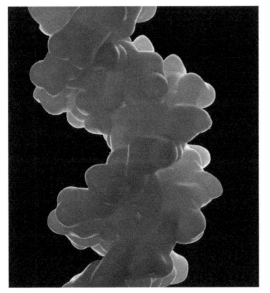
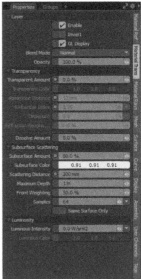

FIGURE 6.5 From top to bottom we increased the subsurface scattering amount

of the shader leaving all other parameters set as default.

In the top image, the aspect of the object (a surface obtained from a DNA sequence) has a plastic aspect. Increasing the reflectivity confers on the object a more milky/waxy appearance.

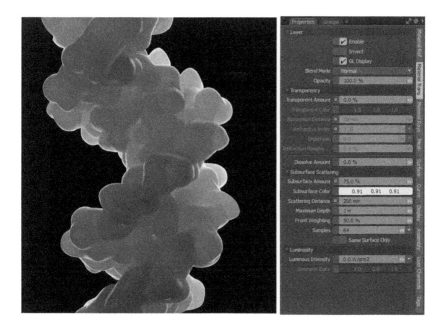

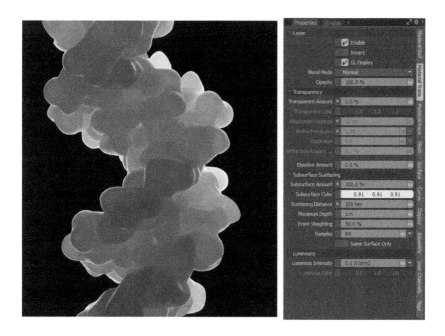

FIGURE 6.5   (continued).

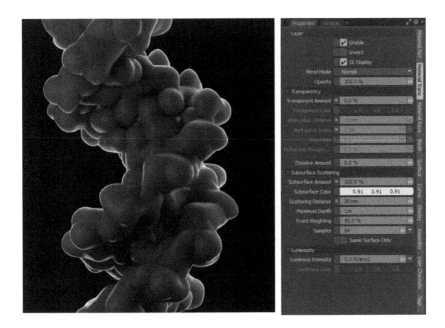

FIGURE 6.5 (continued).

CHOOSING SHADERS

We already presented how important it is to consider the audience and the message that you would like to convey with your image. Colors affect the perception of the subject and the shaders. Visual metaphors should be used wisely in order not to shadow your message. As an example, we will focus on the depiction of an NaCl crystal. What are the different messages that the depictions presented in Figure 6.6 give to the audience?

There are a few aspects upon which I would like to remark. The color of an object depends on the interaction of light AND matter. One of the most common misunderstandings while starting to work with shaders is that the overall color of an object depends only on the diffuse channel, as in Figure 6.7, where we changed only the light setup.

A good knowledge of photography can help us in the rendering of our scene. It is not easy to take *eye-catching* images of mirrors, specular surfaces, and waxy objects without keeping in mind that

FIGURE 6.6 Top left. The main focus is the compact structure of NaCl thanks also to the absence of colors. Top right. In this variant, the focus is on the assembly of the different elements. Middle left. Same as the top right, but in this case the model is a mockup of real molecular model (the plastic model used for teaching in the classroom). Middle right, same as middle left even if leaving out the colors adds focus to the packing of the unit cell. It is still the worst choice for depiction since it overcomplicates the simple message given in the top left image. Bottom images. Two variations of a photoreal rendering of a plastic model mockup. The message is the same as in a picture of the classroom plastic model.

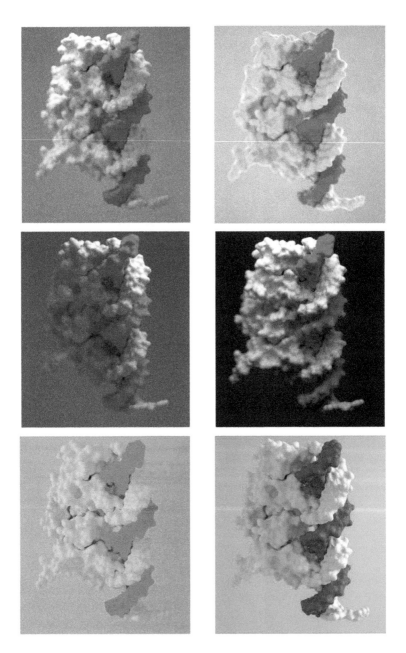

FIGURE 6.7 Renderings made using lighting scene available at flippednormals. com. ⓜ
Shaders are the same for each scene represented.

1) even if not visible on the camera, environments cast light and will be reflected by the shaders applied to your material. If you are going to use mirrors or very reflective material, you have to keep it in mind (see Figure 6.6). The bottom images are made using the same shaders and changing the environment. Reflections (visible on the surface of materials) drastically change the final aspect of the image.

2) When you use subsurface scattering, you should use a lighting setup that can enhance this behavior. The best way is to light your object from behind and use a dimmed light in front (square lights that imitate panel lights work best).

3) If you are going to use translucency effects, you will need a carefully planned lighting setup. I suggest using the same studio setup used by photographers while taking pictures of glass. Lateral panels, reflectors, and a wide selection of backgrounds are indispensable.

A FEW MORE TIPS ABOUT MATERIALS

The more complex the behavior of light is that you are trying to render, the longer the time your render engine will take to compute it. Caustics (those patterns that you see every time light passes through glass), subsurface scattering (skin, waxy materials), chromatic aberrations (shifts of colors when light passes through glass), layered materials, etc., will all increase render times. As we have seen about color, we need to choose shaders according to the information we want to share with the readers/viewers. In my workflow, I use the following approach:

1) When I need to create illustrations for diagrams, I prefer to use a neutral material with almost all of its aspect due to its diffuse channel. The other settings are low reflection, no refractions, no coating layers, no subsurface scatterings, and a low value of roughness for the diffuse channel (see Example 1, Figure 6.8).

2) When I need to create a reproduction of an atomic microscope, I use a self-illuminating material obtained by applying a Fresnel gradient to the self-illuminating channel. This is true for almost

any render engine embedded in 3D mainstream software. Shaders created in this way need a very easy lighting setup (sometimes it is not even necessary to illuminate the scene as the light from the objects is enough to light the whole scene), and the results are very similar to the ground truth (see Example 2, Figure 6.12).

3) I use a lot of subsurface scattering since the lighting setup is simpler than the setup of a scene where there are very reflective/refractive objects. The final aspect is similar to what we see while using electronic microscopy (since you do not see reflections or transparency while capturing images with an electronic microscope) while giving good control to the artistic aspect of the overall image (see Example 3).

4) I avoid extreme values for each parameter. I do not use total black or white, total transparency or reflection, etc. The image looks too artificial while using these values.

5) I add a bit of noise (perlin) to the diffuse channel to avoid creating images that look too artificial. You can experiment with adding noise to reflection/refraction channels to give a better-looking aspect to your shaders. Scratches and dirt are present at a microscopic scale.

STEPS OF MY WORKFLOW

1) Create an overall wireframe of the image in order to give an idea of the model complexity and total number of polygons in the scene.

2) Choose the kind of lights used and start are positioning in the virtual stage.

3) Setup of camera: choose which kind of camera is more fit for the scene. Should I use a wide angle to create a dynamic scene or do my image will need to be a macro shot and use a tele lens with a shallow depth of field?

4) Light settings:

a. I start to target one light at time and then choose its color.

b. I test combinations of ratio of the intensity of the lights used in order to build contrast in the scene.

5) Shaders used in the scene (the core subject of this chapter).

a. Even if you will do not use the same software, you can see how the values of the different channels are mixed together. With minor adjustments, you should be able to reproduce the same shader in your rendering thanks to the principle shaders used.

b. I start from the diffuse colors of all the object in the scene in order to follow a previously chosen palette

6) Render settings. I start creating a quick render with draft settings to see "how" the lights work for the overall scene. Also this let me study what are the zone of the image that look noisier and that eventually require attention in order to increase the "quality" settings for my lights.

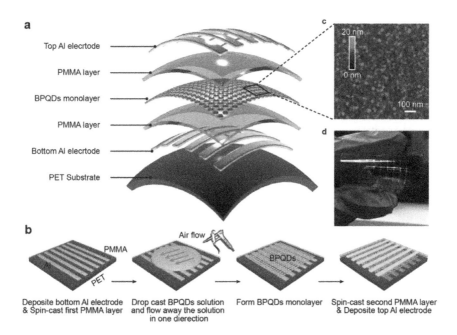

FIGURE 6.8   Figure 1 from the article by S. Han, L. Hu and X. Wang et al.

Figure 6.8 is inspired by Han, Hu, Wang, et al. 2017. "Black Phosphorus Quantum Dots with Tunable Memory Properties and Multilevel Resistive Switching Characteristics." *Advanced Science* 4 (8). Published under CC 4.0 and is based on a figure that represents the schematic diagram depicting the basic fabrication process of flexible BPQD (Black Phosphorus Quantum Dots) based RRAMs.

As can be seem in Figure 6.9, all components of the assembly excluding the BPQDs layer were modeled using box modeling and then an FFD modifier in order to obtain a "bended" appearance. The BPQS layer was modeled bleveling a single cylindrical element and then instancing it with a mograph instancer (in Cinema4D). All instances were then connected and merged with a plane and modified with FFD.

The setup of the scene reproduces a photographic studio: two lateral panels and one curved background in order to avoid harsh shadows in the image. The studio was also modeled using box modeling. The panels are the faces of a box while the background was modeled bleveling the edge connecting two faces of a cube.

As shown in Figures 6.9 and 6.10, I used a classic studio setup. A left warm and right cold panel with an intensity ratio of 2:1, a low intensity front panel and then a back and top panel with the same intensities. All of the lights have the same exposure except for the left warm panel, which is slightly higher. Also, I needed to increase the intensity of each light (as can be seen in the small orange part in the bar of intensity due to the overall scene scale). Since the units in this image are arbitrary, I preferred to tweak the intensity of light while I was free to model my scene without thinking about unit constraints. In Figure 6.10 2, the camera setting is reported: A classic 50 mm camera with 35mm sensor size and white balance set at 6500K. All other settings are the default.

All of the shaders are very simple to minimize the rendering speed and to avoid creating a fake realism that can distract the audience. There is a diffuse channel with low roughness and almost no specularity. The transparent material has an index of refraction of a value comparable with glass/plastic, and almost all settings are set to the default glass from SolidAngle Arnold standard library. I avoided

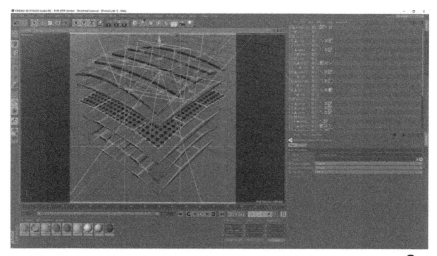

a

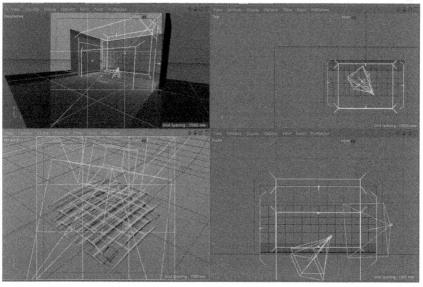

b

FIGURE 6.9 Wireframes of the scene.

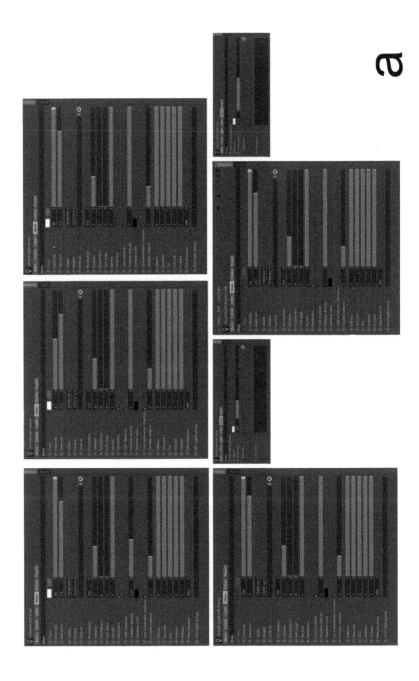

FIGURE 6.10   Lights and Camera.

 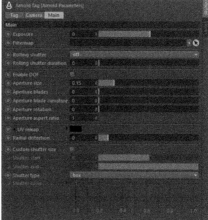

b

FIGURE 6.10 (continued).

adding abbe effects and changing any other parameter as it would have created a non-realistic aspect.

Render settings (Figure 6.11 2) were the following:

- Camera AA: 4 was the minimum needed to keep the image from looking too grainy.

- Diffuse 2: again the minimum needed to keep the image from becoming grainy

- Specular 1: I have almost no reflection in the scene (since also I used a very simple glass) so it was not useful to increase it.

- Transmission: 4 was the minimum value needed to obtain a transparent glass.

- SSS 1: Subsurface scattering materials are not present in the scene.

- Ray depth, I increased the transmission settings while keeping all others low.

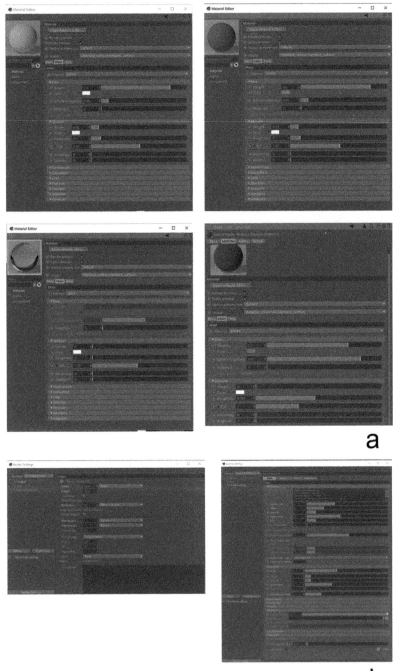

FIGURE 6.11 Shaders.

The final image is not without noise but is fine for a schematic. Arnold rendering does not need to set up ambient occlusion (related to the self-shadowing of objects), but if you use another kind of rendering that needs to specify the use of ambient occlusion, I would turn it on (In the Cinema4D physical render engine, you need to add it as a separate rendering option).

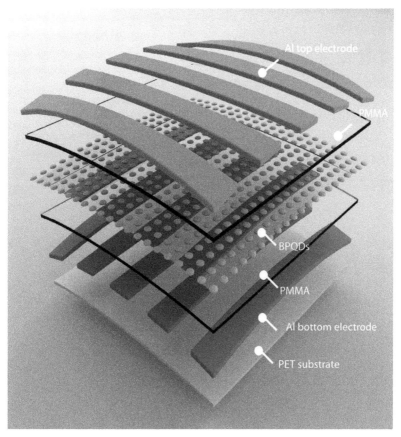

FIGURE 6.12 Final result.

EXAMPLE 2 FAKE ELECTRONIC MICROSCOPE IMAGE

Inspired by several electronic microscope images of white and red blood cells that can be found at cellimagelibrary.org.

The modeling of this scene is quite simple. Red blood cells are subdivided into deformed cubes as seen in Figure 6.14.

For the red blood cells, the steps are the following:

a) Create a cube,

b) Subdivide,

c) apply a deformer (bulge in Cinema4D),

d) select polygons using soft selection, and move them.

For the white blood cells:

e) Create a cube and a small sphere,

f) subdivide the cube,

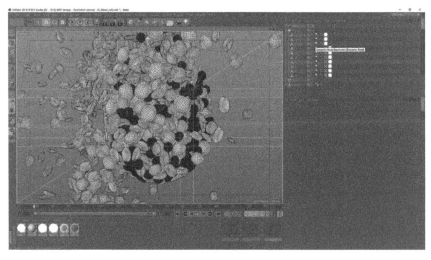

FIGURE 6.13   Wireframes of the scene.

g) instance the small sphere on the surface of the deformed cube, and

h) eventually remesh the result (all software presented in this text has specialized tools for remeshing).

In order to create a *clot*, I created a small simulation. I filled a metaball (created using two spheres) with instances of the modeled objects. The dynamic tags took care of simulating Newtonian behavior for the blood cells.

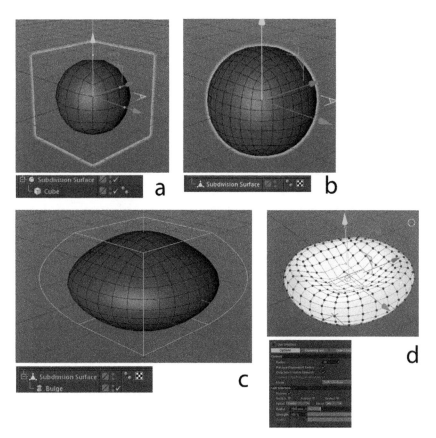

FIGURE 6.14   Modelling of the scene.

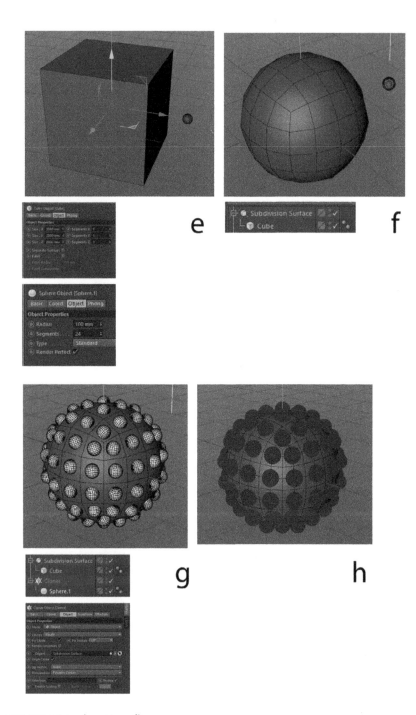

FIGURE 6.14 (continued).

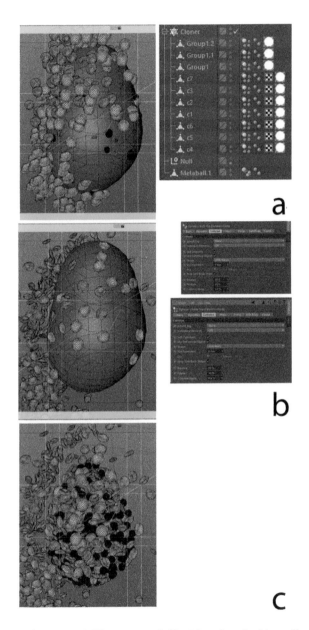

FIGURE 6.15 Clot created filling a metaball with red and white cells.

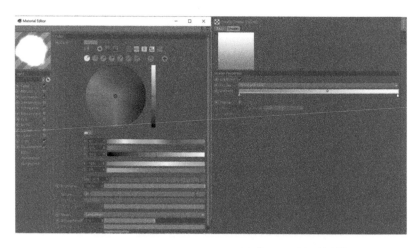

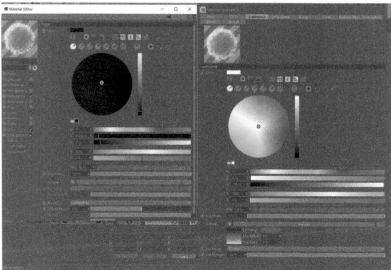

FIGURE 6.16   Shaders settings.

Lights and Camera

The setup is very easy since in this special case the shaders used create the light we need for the rendering. The camera is a 50 mm lens camera with default settings.

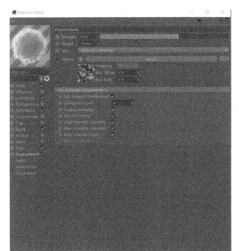
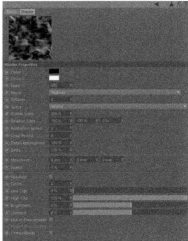

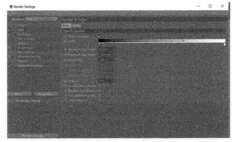

Final Result

FIGURE 6.17   Rendering settings.

Shaders

There is only one shader in the scene. It is a self-illuminating material (the luminance channel takes care of it). Instead of using a plain color, I used a fresnel gradient. This step is important since in this way we will have a variation in the luminosity of the objects according to their shapes. In order to obtain a bit of variation in the shader and in the shapes of the cells, I also added noise maps in the bump and displacement channel.

Rendering

The render engine used is the Cinema4D physical renderer. Default light are turned off. Also, to enhance the details of the objects, I added the effect ambient occlusion. Since all of the lighting is created by the objects, the computation of the final image is very fast.

EXAMPLE 3 ARTISTIC INTERPRETATION OF A MEMBRANE

Figure 6.21 is an artistic interpretation of a cell membrane. The references for the image were taken from the educational poster session on pdb.org to maintain a realistic concept of a cell membrane.

With the aid of Chimera, I imported the pdb files and created surfaces that were then exported and remeshed in Cinema4D. The composition was created by instancing the lipids on a modified plane. The transport proteins were inserted in the holes created in the plane to avoid contact with the lipid bilayer.

The camera used was 50 mm camera using a shallow depth of field. Four lights were present in the scene: a pair of warm and cold lights focused on the center of the image; one light enhanced the subsurface scattering; and a front light with low intensity gave an overall uniform lighting of the scene. All lights cast volumetric shadows to convey the idea of a liquid environment.

There were only two shaders in the scene, both SSS shaders. As seen in Figure 6.20 they did not respect the energy conservation principle as I boosted the SSS contribution. To create a bit of variation in the image (already present due to the different behavior of the SSS material due to the angles of the incident lights) I added a random hue modifier to all the instances in the scene. Also, I used the randomwalk option to create an

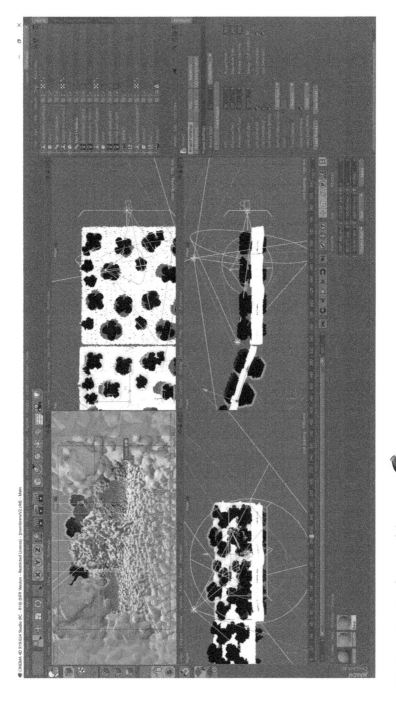

FIGURE 6.18 Wireframes of the scene.

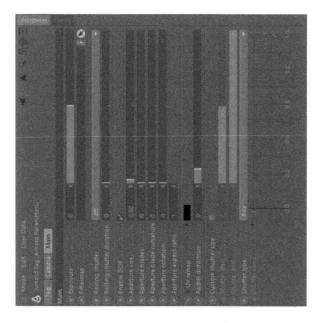

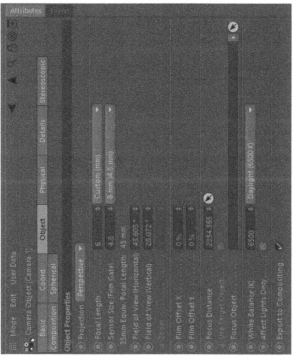

FIGURE 6.18 (continued).

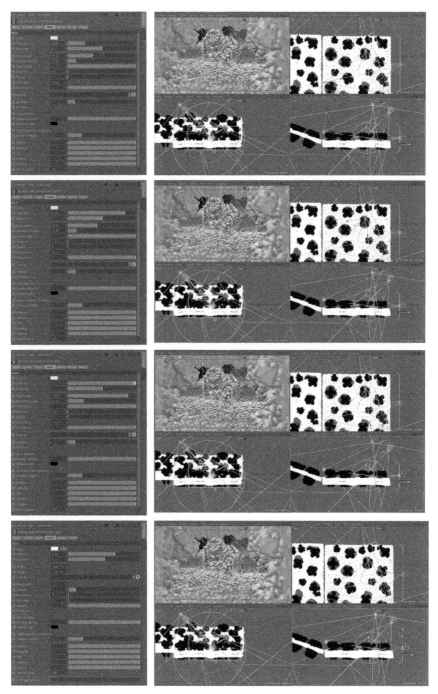

FIGURE 6.19 Lights and Camera.

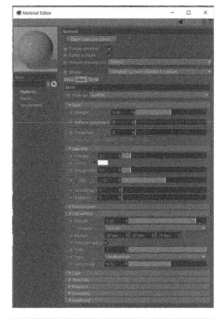
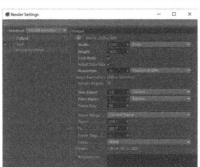

FIGURE 6.20   Shaders.

FIGURE 6.21   Final Rendering.
Top original raw image, Bottom contrasted image.

improved SSS look in comparison with the "traditional" algorithm used by Arnold Render.

Shader settings were the following:

- Camera AA: 6 was the minimum to keep the image from looking too grainy. Since we used SSS, the minimum value is higher in comparison with the previous scene where we used a simpler shader.

- Diffuse 2: Again the minimum to keep the image from looking grainy

- Specular 2: I have almost no reflection in the scene (since also I used a very simple glass), so it was not useful to increase it.

- Transmission 2: a low default value; despite SSS (the main protagonist of this scene and related to Transmission), there is a specific setting that takes care of its quality.

- SSS 4: Since all the scene used SSS, I increased this value in comparison with the other

- Volume: indirect 2 since all lights cast volumetric shadows

Ray depths are keep at a minimum. I increased the transmission value in comparison with the default and lowered the volume depth. This means that even if the volume samples are high, I will skip calculating the indirect bouncing of light in an effort to keep the render times acceptable.

The final image looked a bit flat, so this time I corrected levels, brightness, and contrast using Foundry NukeFX.

UV MAPPING

After getting acquainted with shaders, when the complexity of the models you create increases, you will notice that it can be useful to define the properties of a shader for each face of your object (or portion of it). Even if you are changing your modeling workflow, splitting your objects into smaller parts can be a good strategy to achieve your goals; eventually you will need to map the behavior of your material.

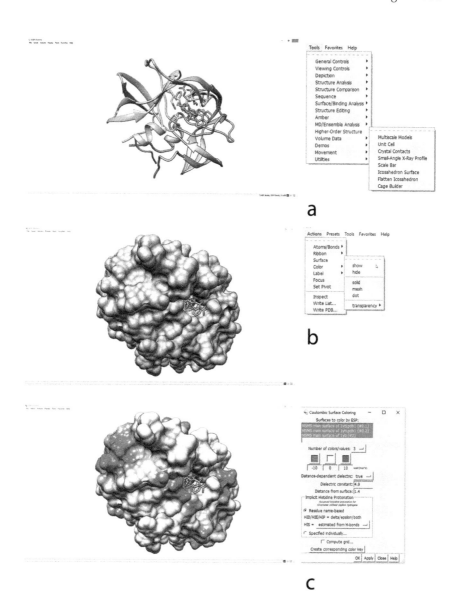

FIGURE 6.22   After importing the model in Chimera, we create a high surface
representation of the molecule a) and b) creation of the surface
and calculations of the coulombic surface information represented using a red, white and blue gradient.

The information is automatically stored in the vertex of the dae file.

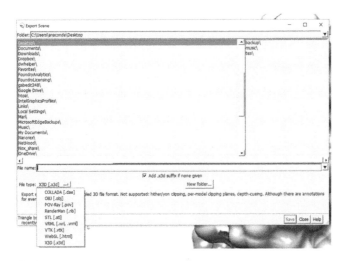

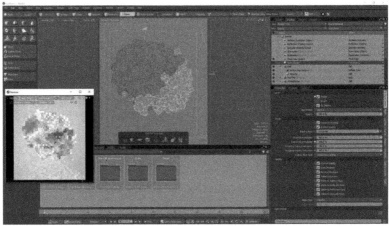

FIGURES 6.23 AND 6.24 The info present in the vertices of the model is projected on a diffuse texture map created after automatically unwrapping the mode with the tools present in foundry MODO.

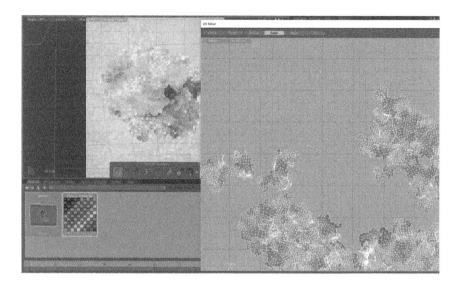

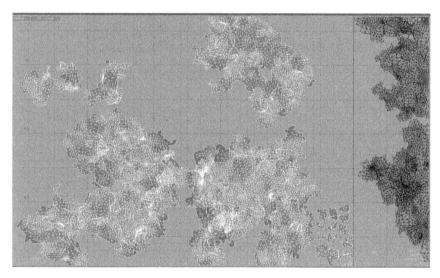

FIGURES 6.23 AND 6.24   (continued).

FIGURE 6.25   Final result.

UV mapping and textures are helpful. We have already talked about UV and texture mapping. We can think about them as stickers that are linked to the surface of our models and can save information about them. UV maps and textures are indispensable when we need to bind colors to the spatial physical properties of the molecules. For example, we can see the representation of a hydration surface of a molecule. Here, colors are not the expression of an imaginary world but are used to depict a specific property of the molecule in a selected area.

In figure 22 et seq. example, we imported a molecule from PDB databank, created a surface, and applied colors to a vertex after performing a calculation. We can then import the molecule into Modo, which can take care of creating a UV map and a linked texture used in the diffuse map of the shader. As we can see, if necessary, we can directly paint on the texture in order to change it. If no variation is needed, we can render the model without changing the texture, obtaining a high quality version of the Chimera viewport.

When working with UV generally, we think about diffuse channels, but we are not limited to this. We can combine maps in order to experiment with very interesting results. Since our work generally involves the use of a lot of objects, we do not need to manually paint the map for each object and the workflow; we make use of a large number of auto-generated UV coordinates.

BUMP, DISPLACEMENT, AND NORMAL MAPS

When representing nanomachineries, we will generally not make use of *bump maps* or *displacement* maps, but we should know about their use. We have seen that there are different parameters for aspects of a shader. We have presented the *principled shader* paradigm that covers almost all the phenomena that happen when rays of light interact with matter.

Is it possible to use a map to reduce the polycount of our objects and fake their shapes? While depicting the aspect of a novel engineered biomaterial, can we make use of the information gathered by an atomic force microscope or electronic microscope? The answers are, of course, yes. Special maps such as bump, displacement, and normal maps can take care of this task as all of them simulate the interaction of light with the asperity of the surface of the models faking its geometry.

How do they differ? *Bump maps* fake the bumpiness of the surface *without* actually changing the geometry of your object (i.e., the shadows will be the same as the unmapped object).

Normal maps contain information on the angle of the normal of surface of your object. They are similar to bump maps. If you use a real-time engine for rendering, you will rely on them.

Bump and normal maps are almost equal in their use of resources.

Displacement maps change the geometry of the object (in this case, the shadow of the object will not be the same as the unmapped object). They require more resources than bump and normal maps, but the final result is of better quality. Each software included a tool for generating and mixing basic noise. In Figure 6.26, I reported the noises I use and a few (the 4 bottom images) examples for creating special maps.

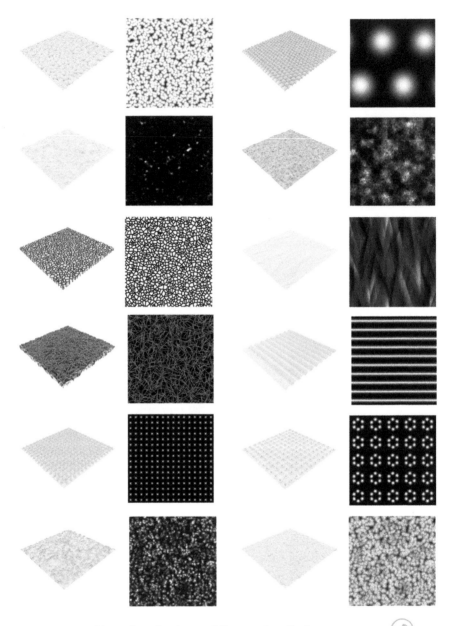

FIGURE 6.26   Examples of noise used for creating displacement maps.

In the example scene, we used a plane; the software took care of applying the best UV coordinates. As we have seen, we can create customer UV texture in order to set up the position where we would like to use displacements.

HOW A RENDER ENGINE CAN CALCULATE THE FINAL ASPECT OF YOUR IMAGES

We have used most default render settings and employed custom settings in the previous example to minimize render time and optimize the quality of the images. If we would like to keep the analogy of a render engine with a camera, we can say that using a universal setting is quite similar to switching your camera to auto. It can do a great job, but it is also interesting and useful to know how to tweak the settings according to the images we are rendering. Also, we can create presets (analogous to the programs in our camera) to spare time in rendering these images, but we need to answer the question of what a render engine is.

Most people have seen a movie from Pixar. I remember reading in the CGSociety forum that a simple way to answer the question: "What is your work, and how do you use a computer for drawing?" is "I create images as you can see in the movies of Pixar." Pixar, in their open course about 3D graphics released at Khanacademy.org, reports one of the most clear and concise explanation of how a render engine can calculate your images:

> Rendering is the final process of a movie, when the technical directors at Pixar calculate the color of every pixel in every frame of every shot in the film. If that sounds incredibly time-consuming, it is! But Pixar gets help from some mathematical equations.

Every render engine, in one way or another, tries to solve *the* rendering equation to calculate the color (and opacity) of each pixel that will represent a frame of the movies you will see. The rendering equation is the following:

$$L(x \to \omega_0) = L_e(x \to \omega_0) + \int_{\Omega} L(x \leftarrow \omega_i)\rho(x, \omega_0, \omega_i)(n_x \cdot \omega_i),$$

For any surface in the scene, the outgoing radiance $L(x \to \omega_0)$ leaving a point x in a direction ω_0, can be described as the sum of the emitted radiance $L_e(x \to \omega_0)$ and the reflected radiance $L_e(x \to \omega_0)$ at x toward ω_0 (assuming that there is no participating media in the scene). Ω is the visible hemisphere. $\rho(x, \omega_0, \omega_i)$ is the bidirectional reflectance distribution (BRDF) describing the surface reflection.

Even with very fast computers, a theoretical approach is not suitable for simulating every single photon of light that interacts with the surface of an object. We need to use a shortcut. We have already seen how we can fake a reflected light, adding point lights around our object where we expect the light to be diffused or reflected by the objects in our scene. This time we will cheat to calculate our equation more quickly.

We can consider each object as a source of light and calculate the interaction with its environment (indirect methods). We can calculate how the object creates self-shadows (occlusion); we can precalculate and cache where the rays will converge (lighting cache, final gathers). We can reduce the number of rays to trace using stochastic methods and choosing a priori how many reflection, refraction, and SSS rays to trace. We can de-noise the image obtained using several algorithms (filtering a noisy signal).

The more your render follows the coded laws of physics while solving the render equation, the more it will be considered an unbiased physical render engine. Otherwise, you will rely on a biased render engine. This definition is sometimes marketing hype. Actually, every render engine needs to take some shortcut while solving the rendering equation. There is no agreement on the terminology used in CG. Definitions of terms like global illumination and direct and indirect lighting can be tricky.

The concepts I would like you to keep in mind while setting up your render are

- We are using approximations for solving an equation, so it is quite common to find render engines that let you choose different algorithms to complete this task.

- Some methods are more rooted in the physical law that governs the behavior of light while interacting with materials. This does not necessary means that you will obtain the best results!

- All the tools we use should be actual tools and not the focus of our representation.

Now we will look at the options offered by our software, focusing on the common settings.

All of them include

- Format of the final image (or the single frame); when possible use an uncompressed file format with the highest resolution

- Anti-aliasing; all of the algorithms found under this setting refer to the final noise in your image. Some algorithms work better for images with a lot of contrast while others are used for universal settings. Generally, software offers the choice to use adaptive anti-aliasing setting. This means that the software will try to calculate where you will need more samples to avoid artifacts in your image. This can take longer than a fixed sample rate.

- Samples setup (i.e., reflection samples, refraction samples, subsurface samples). Samples are related to the quality of your render and the number of rays shot in order to solve the rendering equation. Increasing the overall quality will increase render times.

- Ray depths or limits (i.e., reflection depth, refraction depth, reflection limit, refraction limit, etc.) are related to the number of bounces your rays will perform before they no longer contribute to the calculation of the scene. These refer to the specific interaction of light with your subject. The higher, the better, but as we have seen for the samples setup; increasing them will also increase the render times.

In the Figure 6.27 et seq., I've included a few key examples of how changing these parameters drastically changes the final quality of your image. All renders were created in Foundry MODO.

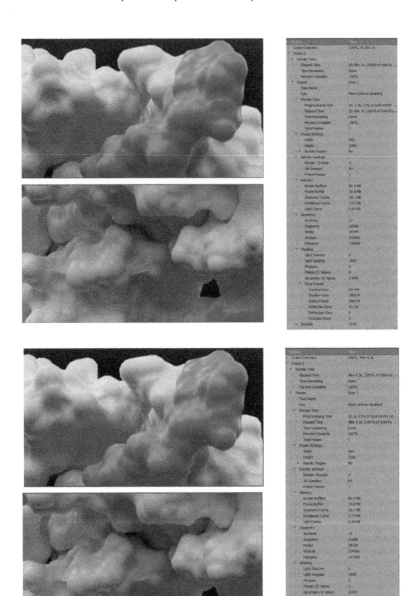

FIGURES 6.27 AND 6.28 Examples of different settings used for the final rendering in a well-lit and dark "occluded" part of the image.

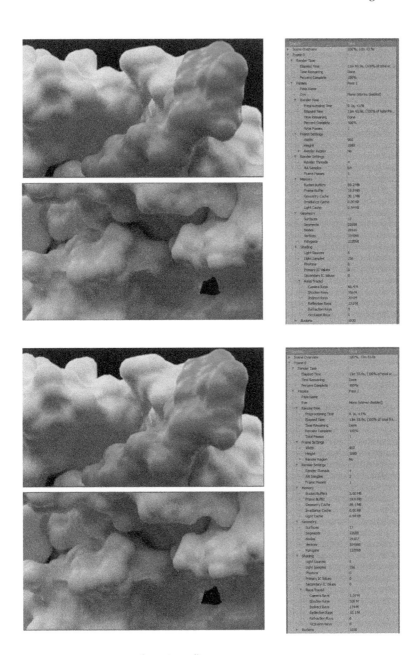

FIGURES 6.27 AND 6.28 (continued).

Render times vary from a few minutes to a few hours for just one image. The noise looks quite different if we check an area with a lot compared to a surface where there is plenty incident light and fewer asperities that can trap light. In the first case, we would need a more accurate integration since the shape of the object will trap the light and the render engine will need to calculate more bounces of the light rays to create a cleaner image.

I want to conclude this chapter with the notion that you have to create images that can stand the test of time. Do you remember when everywhere you saw the fake lens flare or fractals were the subject of every CG movie? It did not seem overdue but now it looks cliché and overdone. Your goal is to create a high quality image that uses the best tools to represent a concept and not to showcase the latest feature of your software.

EXERCISE

- A good exercise is to change the appearance of an object simply by changing the color of lights. Try to open one of your scenes and give it a totally different appearance using only the lighting.

- Try the same exercise as in the previous point but this time change only the shader.

- Try to find at least a few representations that can be a challenge to render due to the number of objects in the scene; try to create displacement maps that can lower the polygon count and then compare the rendering times of the two images.

- How far can you push the modeling while using only displacement maps? Focus on the search for electronic microscope images; while building a self-illuminating shader, try to use displacement maps to add noise and details.

- In order to optimize the render settings of your scene, try the following. Analyze the shaders used in the scene. Which interaction between matter and light simulates most? Do you need the same depth for all behaviors generally calculated by integrators?

- Although there are plenty of default material libraries, create a small personal shader library trying to include your personal fake electronic microscope material and subsurface default material. Try to apply them to different models and see how they are influenced by the lighting conditions and by the model topology. In this way, you will understand how they work. Sometimes if you reinvent the wheel, you can understand how a wheel works.

SUGGESTED READING

Birn, J. 2015. "Lighting and Rendering." In *Lighting and Rendering*. New Raiders.

Dal Corso, Alessandro, Jeppe Revall Frisvad, Jesper Mosegaard, and J. Andreas Bærentzen. 2017. "Interactive Directional Subsurface Scattering and Transport of Emergent Light." *Visual Computer* 33: 371. doi:10.1007/s00371-016-1207-2.

Firbank, M., M. Hiraoka, M. Essenpreis, and D. T. Delpy. 1993. "Measurement of the Optical Properties of the Skull in the Wavelength Range 650-950 Nm." *Physics in Medicine and Biology* 38 (4): 503–510. doi:10.1088/0031-9155/38/4/002.

Goodsell, David S. 2012. "Illustrating the Machinery of Life: Viruses." *Biochemistry and Molecular Biology Education* 40 (5): 291–96. doi:10.1002/bmb.20636.

Han, Su Ting, Liang Hu, Xiandi Wang, Ye Zhou, Yu Jia Zeng, Shuangchen Ruan, Caofeng Pan, and Zhengchun Peng. 2017. "Black Phosphorus Quantum Dots with Tunable Memory Properties and Multilevel Resistive Switching Characteristics." *Advanced Science* 4 (8). doi:10.1002/advs.201600435.

Jakob, Wenzel. 2014. *Mitsuba Documentation*.

Jakob, Wenzel, Eugene D'Eon, Otto Jakob, Steve Marschner 2014. "A Comprehensive Framework for Rendering Layered Materials." *ACM Transactions on Graphics* 1–64. doi:10.1145/2601097.2601139.

King, Alan, Christopher Kulla, Alejandro Conty, and Marcos Fajardo. 2013. "BSSRDF Importance Sampling." *ACM SIGGRAPH 2013 Talks on - SIGGRAPH '13*, 1. doi:10.1145/2504459.2504520.

Kwast, Daniël Jimenez. 2014. "An Introduction to BRDF Models." *Hmi.Ewi. Utwente.Nl*. http://hmi.ewi.utwente.nl/verslagen/capita-selecta/CS-Jimenez-Kwast-Daniel.pdf.

McDermott, Wes, and Allegorithmic. 2012. "PBR Shading." In *2012 IEEE 10th International Conference on the Properties and Applications of Dielectric Materials*, 2: 1–11. IEEE. doi:10.1109/ICPADM.2012.6318948.

CHAPTER 7

Animation

PRINCIPLES OF ANIMATION

As we have seen in previous chapters, when you set up your 3D environment, you are recreating something very similar to a photographic studio. You set up lights and cameras, "pose" your model, and then render the image. As rendering static images is analogous to taking photos, animation is analogous to movie making. There are useful tutorials and books dedicated to this topic; this one chapter cannot be exhaustive. We will focus on introducing the most common techniques for representing nanomachineries. Before starting to analyze them in detail, we need to discern whether there are differences in the principles that govern animating molecules and those that apply to traditional animation of simple 3D objects.

The principles are still the same from the days of the zoetrope (this is also true for AR and VR!); we need to create a sufficient number of frames (eventually in real time) for each second of animation (starting from 24 frame per second). We can create *key* frames and let the software calculate the missing ones, or we can manually create each frame. At the end, we will merge all the images and create a movie (we will spin our zoetrope!). In other words, no matter how advanced the tool used to produce our movie, we will have to deal with creating a sufficient number of frames.

We will organize this chapter with the same principles used before. We need to know the audience of our work in order to know how scientifically accurate to make our animations. We can range from the actual simulation of biochemical processes to the artistic interpretation of them.

The first category (simulation based) includes protein docking, changes in the conformation of molecules, wiggling molecules, chain unfolding, chain building, etc., while the second one (artistic interpretations) include floating molecules, macroscopic destruction of proteins, building and swarming of simple assemblies of molecules, etc., according to a simplified version of Newtonian physics (i.e., all the particles will interact as small spheres without considering the interactions at an atomic level).

Simulation Based Animation

All the specialty software we have seen so far has interfaces with the most used molecular modeling software and routines that let it accomplish very specialized tasks (i.e., docking molecules, moving according to rigid constraints given by physicochemical laws, wiggle atoms, simulating several kind of molecule motions that are related to their spectra, etc.).

Chimera is a type of this software, and among the different tasks it can accomplish, it can create animations of

1. rocking bonds, playing a molecular trajectory, morphing between density maps, waving morph quaternary structures, wiggling molecules, and backbone tracing,

just to name a few. Chimera includes a command that can export the status of our molecules (position, attributes, maps, etc.) for each frame of the animation.

The models exported (in obj, dae, x3d, among other formats) will work as in a stop-motion animation where a model state is modified for each frame to create the final movie.

Example 1 Morphing Molecule

In order to create animations, Chimera uses a scripting language with very intuitive commands. Thanks to very detailed documentation, we have presets that can achieve almost all the animation we will need. Let us start analyzing one of the scripts in the Examples section of the documentation of Chimera, one that creates a morphing animation among different states of the same molecule (pdb id:1r7w).

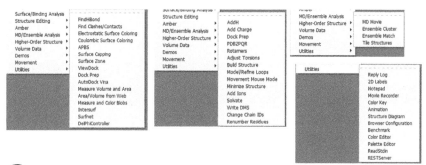

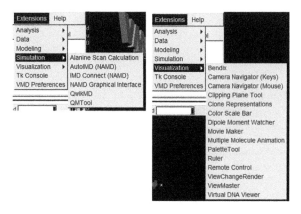

FIGURE 7.1   a) Chimera. b) VMD menus dedicated to animations.

```
open 1r7w
turn x 120 ; scale 1.2
morph start #0.1
morph interp #0.2 ; morph interp #0.3 ; morph interp #0.4 ; morph
interp #0.5
morph interp #0.6 ; morph interp #0.7 ; morph interp #0.8 ; morph
interp #0.1
morph movie
~modeldisp #0
```

```
movie record supersample 3
coordset #1 1,161
wait 160
movie encode coordset.mp4 coordset.webm coordset.ogv bitrate 200
```

We can translate the script in pseudocode as:

```
open the molecule,
turn the molecule 120 degrees
start the morph of the model called 0.1 using present density
choose the interval for the morph interpolation
display only one model
set up the render settings
record the movie
```

Now let us see what happens when we copy and paste this command to the command line console present in Chimera.

```
If we prefer to see a volume morphing, we need to replace the morph
command with molmap and then vop.
molmap #0.1 3
molmap #0.20 3
vop resample #0.20 onGrid #0.1
vop morph #0.1,1 playRange 0,1
```

These are the instructions we gave Chimera translated in pseudocode.

```
create the molecular surface of model 0.1 with resolution 3
create the molecular surface of model 0.20 with resolution 3
resample model 0.20 using the grid settings of model 0.1
create the morph
```

After executing the script, the animation will play in the dedicated movie player. In order to set up the artistic aspect of the animation (speed, aspect of the molecule, and camera and lighting settings), we can export a model for each of the frame/states in which we are

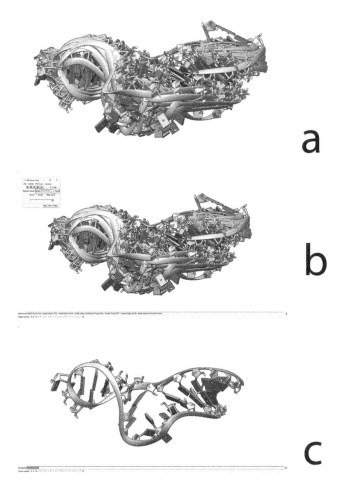

FIGURE 7.2   Importing multiple states of molecules with pdb ID 1r7, rotating it (a and b), and then displaying only model number 0 (#0) (c).

interested, then import them as a sequence of models in Houdini in order to create a model for each state.

In order to do this we can execute the following commands for each frame we want to export:

```
vop morph #0.1,1 playRange 0,0.1
vop morph #0.1,1 playRange 0,0.2
```

and so on....

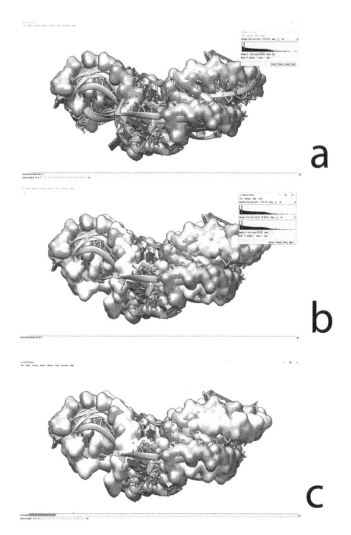

FIGURE 7.3 a) creating a molecular map for model #0 with default grid density of 3. b) creating a molecular map for model #20 with default grid density. c) resampling a molecular map of model 20 with same grid density as model #0. d) deselection of multiple models. e) creation of the final morph.

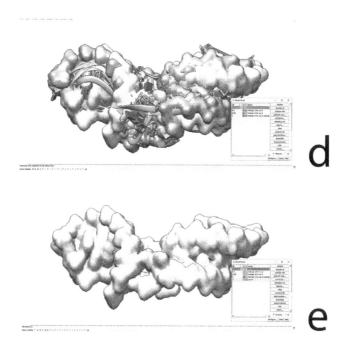

FIGURE 7.3   (Continued).

As we have already stated, specialized software can be the interface to a huge variety of simulation routines/software and consequently also the number (and ways of representing) that we can use. How can we get a clearer idea of how to proceed? A chapter on each animating technique would be tedious, and it would become obsolete quickly.

Again, the answer is to hold firmly to the core concepts of animation; we can summarize in two points:

- Your main 3D software can work with a series of models and create framed animation from them regardless of how complicated the simulation is.

- You can export the models for each frame with all the features (position and shape) and then import them as a sequence in your 3D software in order to set a scene suitable for your movie recording.

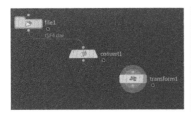

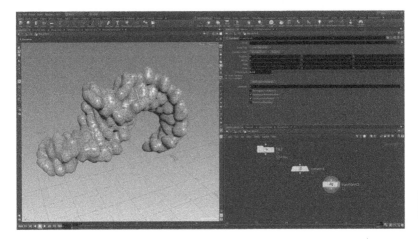

FIGURE 7.4 a) node used for importing the dae sequence in Houdini. b) visualization in the viewport of one of the frames of the animation.

I strongly suggest becoming familiar with a scripting language in order to be able to set up animation in molecular modeling software that you can export to your favorite 3D software.

Non-Realistic Simulation

Very often, we will need to use our imaginations to depict in a more artistic and stylized way nanomachineries' interactions. In these cases, the time spent with the specific 3D scientific software will be shorter while we tweak more aspects of the 3D virtual set where the action takes place. As already seen, we generally export the models from Chimera or VMD and then use the tools present in Cinema4D, Modo, and Houdini. When animations take place on a mesoscopic scale, there is no need for accurate representation of models at atomic levels.

How can we generate molecules or entities to take part in the simulation? One of the most common approaches is based on these principles:

1. All described software has particle emitters.

2. We can force the emitter to emit primitives or any other imported meshes.

After we have created the object needed, we can move it manually, or when needed we can rely on the fact that the particles can be tagged with modifiers in order to follow different rules and that we can stack modifiers in order to create even more complex animations. These principles can be used to simulate all kind of processes, from nanoassemblies to growing fibers, cell division, decay of tissues, constrained growth, etc. etc.

Why Are These Simulations Non-Realistic?
The book *Bionumbers* and the website Scitable (springernature.com) can give you an idea of the small universe(s) confined in our body. DNA replication is a constant process not only for each cell of your body but also for the small machineries (mitochondria) inside it at a pace of 50 bases per second for each nanobuilder in the human body while in the *E.coli*. the speed can increase to 1000 per second.

Although 3D mainstream software can depict fluids emitting a very high number of particles and make them follow a different set of rules (think of how we can perfectly simulate the appearance of the ocean), the millions of elements contained inside one cubic mm of blood are too many to allow for the use of simulation-based molecular modeling.

Moving Objects and Changing Parameters Making Use of Key Frames

Example 2 Self Assembly Inorganic Cell
In the previous section we described how the status of each atom (and molecule) can be simulated, but how can we manually specify the trajectory followed by an atom? As an example, how should we proceed if we want to make a DNA molecule rotate (to create what is called a turntable animation) or move the atoms of an inorganic molecule to create the illusion of self-assembly process?

We will start creating an animation of an NaCl crystal building itself. The main idea is to have each ion that builds the crystalline cage move individually in a straight trajectory from different starting positions. How? Easily:

1) Select the object of interest.

2) Check the *autokeying* options (marked in red in Figure 7.21)

3) Set a *key* and then move to the last key of our animation (in the example presented from the first frame-to-frame 240). Our software needs a reference for the starting position. In this case, this process can be repeated if we need to split our animation into smaller parts.

4) Move the object in the final desired position of each of object you would like to animate (The parameter that can be animated generally gets highlighted; it is generally also possible to animate several parameters at the same time. We are not limited to an object's position; we can focus on their colors, shaders, textures, UV maps, etc.).

After performing these four steps, the software will take care of calculating the parameters for the missing frames. You have to keep in mind that each step will be interpolated *linearly* between the *key frames*. If you want to change this behavior, all software offers advanced editors for setting up animation details (i.e., you want the movement to go faster at the beginning of your animation and then to fade slowly when it approaches the end). This can be done by setting non-linear behavior among the keys.

Example 3 DNA Turntable
We will show an example similar to the previous one, but this time we will rotate the position of the molecule instead of translating it. In the following example, we will create a turntable animation of a small fragment of DNA using the same technique presented in the previous example. In order to better compare the tools that Cinema4D, Modo, and Houdini offer, the example is repeated for all of them
I would like to highlight that

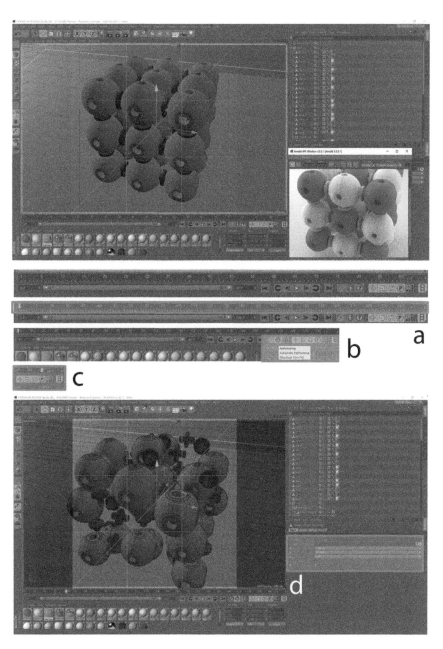

FIGURE 7.5   a) Selected objects in the scene. We changed the frame count from 90 to 240 in order to prepare 10 seconds (at 24 frames per second) of animation. b) Autokeying switched on and setting of first reference key. c) After moving the objects, we set another reference key at position 240. d) The parameters are active (they are now marked in orange), so the software took care of creating all of the frames, and our animation is ready to be played.

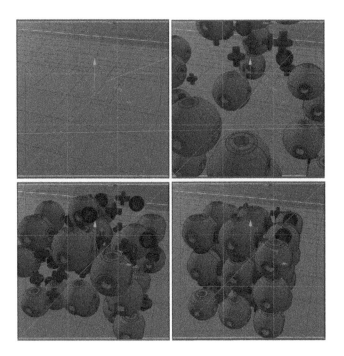

FIGURE 7.6   Steps of the animation played backward to create the illusion of self-assembly.

- Even if we used a visual approach for creating our turntable, we could have scripted each animation by means of proprietary scripting language or Python.

- All software offers the possibility of setting in detail the speed of the animation and changing the linear interpolation for a non-linear one. Figure 7.10 shows an example of how to move a simple object without following a linear interpolation.

Use of Noise

Dealing with molecules means dealing with uncertainty. A good way to communicate this idea is to make use of modifiers that can *add noise* to the coordinates of our objects in time (or when required, to change randomly other attributes such as rotation, properties of the shaders used, etc.). The

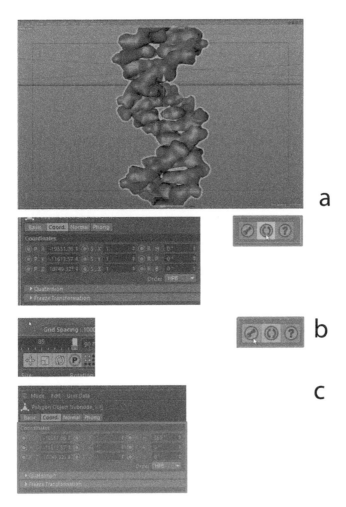

FIGURE 7.7   Turntable of DNA fragment created using Maxon Cinema4D a) setting up autokeying. b) recording the essential key frames. c) the parameters are recorded by the software. Notice the change in appearance of the radio button on the interface.

same is true also for dust particles, corpuscles, pollen, and small cells. Adding noise will keep the eye of the viewer interested. Instead of creating it manually, we can automate this effect by adding a noise modifier as in the following example where a noise modifier is applied to the component of a stylized cell in order to add the feeling that the cell is alive.

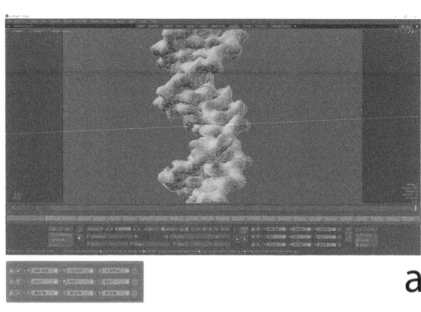

FIGURE 7.8 Turntable of DNA fragment created using Foundry Modo a) selecting the fragment (the contour is highlighted in orange showing that
the software is recording the position of the object; it is set on as
default. b) shifting the timeline to the last frame of the animation, we
change the Y rotation of the molecule. The position is now recorded
as shown by the radio button marked in orange.

Example 4 Living Cell Animation
In this example, we use a sphere surrounded by smaller spheres that
change their positions according to a random noise; all of them are the
components of a metaball that represents the stylized cell.

The steps to create the scene are the following:

1) Create a central sphere.

2) Create smaller clones.

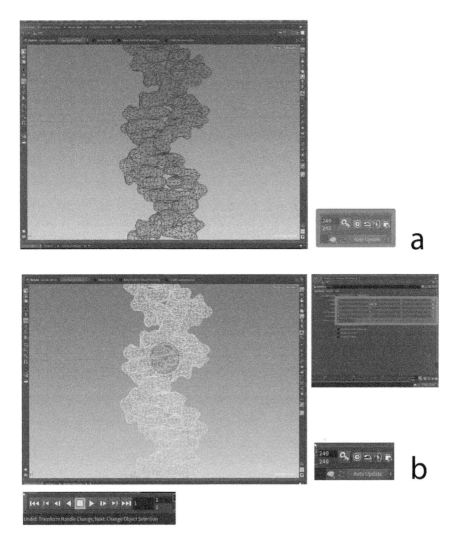

FIGURE 7.9 Turntable of DNA fragment created using SideFX Houdini a) selecting the fragment, activating autokeying. b) shifting the timeline to the last frame of the animation and changing the Y rotation of the molecule. The position is now recorded as shown by the text box colored in orange.

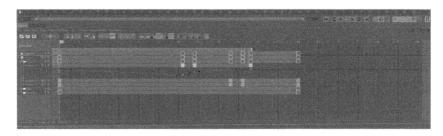

a

b

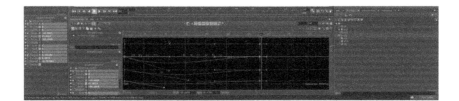

c

FIGURE 7.10   Common tools for creating non-linear variations of the animated
parameters in a) Cinema4D, b) Foundry Modo, and c) Houdini.

3) Merge the central sphere and the clones using a metaball modifier.

4) Add a random modifier linked to the position of the small spheres.

Here we do not need to record the initial and final positions of the spheres since the random modifier will take care of changing them for each frame of the animation.

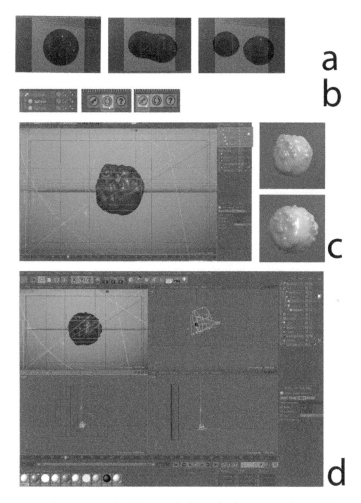

FIGURE 7.11   a) an example of a metaball applied to two spheres. b) a random modifier applied to the position of the small spheres. c) two frames of the animation. d) setup of the scene. One camera, three lights, and a background are present. One light is used to light the background, and two lights are used to light the main cell.

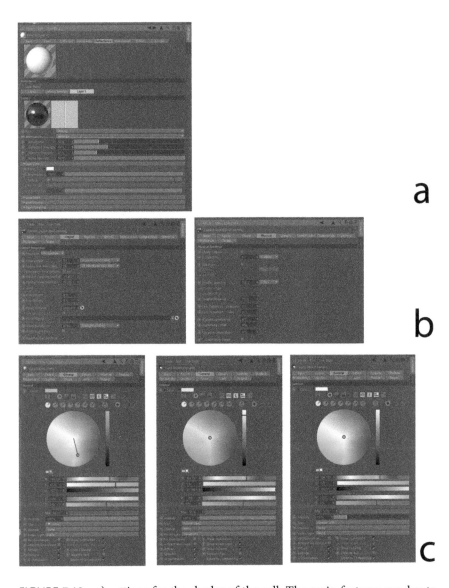

FIGURE 7.12   a) settings for the shader of the cell. The main features are due to the reflectance parameters. An old model (phong) confers a wet appearance (the same parameters are generally used for creating wet surfaces). b) camera settings. 50 mm camera with default settings c) Light settings. A blue light for the background, a neutral key, and fill light for the cells; the intensity ratio for the lights is 100:15.

Particle Simulations

Almost every software can simulate the emission of particles from a source and calculate their interaction. All of them can simulate collisions approximating the object in the scene as solid objects that follow the classic theory of physical impacts. One of the first animations I created was about water (gas) particles in a box.

I remember the first time I was introduced to the study of statistical thermodynamics. The entire universe seemed filled with containers into which we could insert a few molecules at a time in order to study how they translate, rotate, and interact with each other like perfect particles. It required quite a creative effort to depict them and good drawing skills by our teachers.

When I was a tutor, I was very happy that I could show my students a real time simulation relying on the capabilities of 3D software.

How can we recreate this example using the common tools that mainstream 3D software offers?

It is quite easy if we follow these steps

1) Insert a box primitive and edit in order to become a mesh.

2) Import a pdb molecule of water or just recreate it using simple spheres.

3) Insert a particle emitter.

4) Set the particle emitter in order to emit the molecules previously modeled and tag the molecule and the box in order to follow Newtonian rules.

5) Set the way the particles (in these case our molecules) should interact.

6) Start the simulation (we do not need to create keys since the software takes care of doing it while creating simulations).

As seen in the living cell example, we do not need to set up any keys since the emitter will take care of creating all the frames needed.

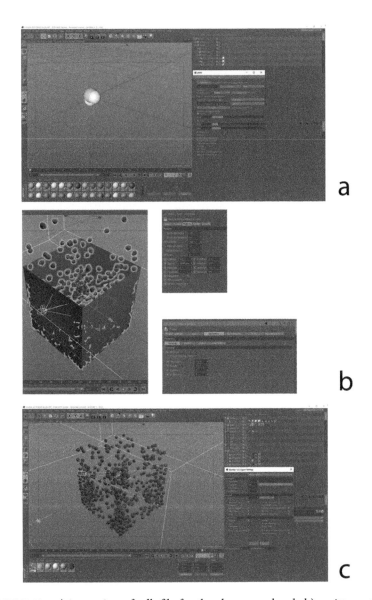

FIGURE 7.13   a) importing of pdb file for the chosen molecule b) emitter settings and project settings. In order to make the particles float, the project gravity was set to 0. c) setting used to export the animation.

Example DNA Assembly

In this example, we will show how to illustrate a stylized DNA assembly process starting from its bases. In order to make the example easier to follow, we will use four different primitive represent the four canonic bases. The idea is to create a small nanoassembler that will get the bases one at time and then create a double helix. Each base will be paired according to specific predefined selection rules.

The following steps are used:

1. Insert the primitives that will mimic the DNA bases.

2. Insert particle emitter.

3. Create the modifiers that will take are of making the bases follow our splines.

4. Create two linear splines that simulate the path of the bases before assembly.

5. Set up the emitter parameters (dimension of the source, speed, rotation, and randomness).

6. Duplicate the emitter and change the order of the primitives in order to be compatible with the other emitter.

7. Start the simulation.

We do not need to set any key. The software will take care of creating the frames.

Example Tracing Particles Trajectory

Another useful tool in most software is particle *tracing*. Using a node or modifier that is able to trace the trajectory of a particle is very useful for creating stripes and fibers in no time. Since we can put obstacles among the path of our particles (and modifiers that can modify their paths) we can wrap the fibers around every kind of shape we need. The following example combines most of the features presented thus far to simulate two fibers growing.

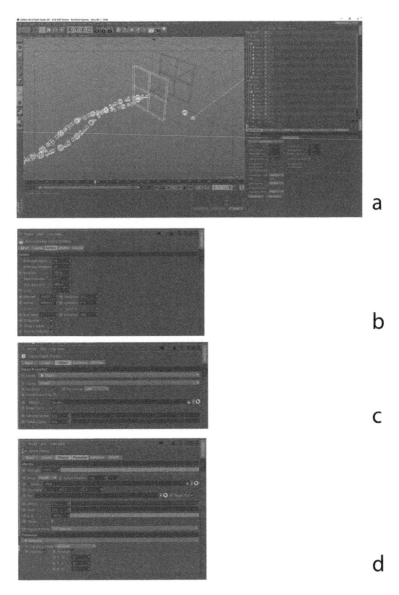

FIGURE 7.14   a) general setup of the scene. We have two strands of DNA formed by a helix, a spline modifier, a cloner, and an emitter. Cinema4D needs to create clones in order to make them follow a path given by the user. b, c, and d) detailed settings of the emitter, the cloner, and the spline modifier.

In order to build this scene in Cinema4D, we need to follow these steps:

1) Create an emitter.

2) Make the emitter use a sphere primitive.

3) Add a modifier to create clones of the spheres.

4) Make the clones follow a path.

5) Trace their path and extrude them.

To create more variation in the fibers we can use the following techniques

1. Copy the previous setup.

2. Instead of the sweep and trace modifier, create a cloner.

3. Make the cloner create spheres.

4. Attach a metaball modifier to the spheres generated by the cloner.

Two More Complicated Examples Using Procedural Modeling

As reported in the modeling section, a new paradigm of modeling is getting more and more attention lately: *procedural modeling*. If you are already familiar with programming, procedural modeling will be a blessing in your work. A reductive definition of procedural modeling includes its capabilities to create instances, clones of meshes, and change their surface/volume according to strict rules coded by the user via coding. A few examples are the growing of a plant, the expanding of roots, cell division, and the growing of a crystal. Since we are modeling them following these rules, we can also animate their growing process making the parameters expressed by the variable of time.

All process presented in nature are time related, so it is easy to find them expressed according to it, which tremendously simplifies our animating tasks. For example, it cannot be too difficult to sculpt a realistic vein/root, but the task of creating a plausible growing animation will require additional time be spent on the scene. When we go the procedural route, the process of modeling and animating are simultaneous.

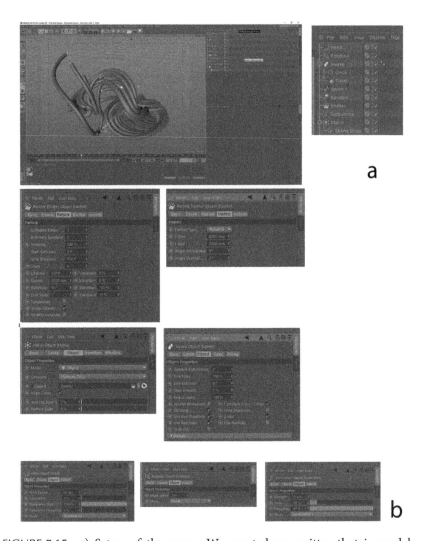

a

b

FIGURE 7.15 a) Setup of the scene. We created an emitter that is used by
the matrix object. The matrix object uses a spline to change
the trajectory of the particles according to a spline. A sweep
modifier is linked to the matrix in order to sweep a circle
according to the tracer modifier. b) Several modifiers (wind,
rotation, and turbulence) add noise to the trajectories of the
particles.

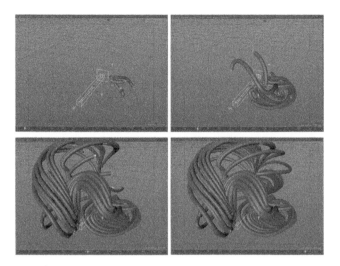

FIGURE 7.15B   Four frames of the animation.

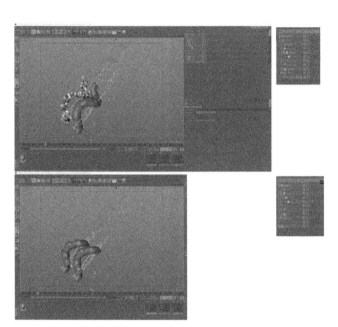

FIGURE 7.16   a) the cloner modifier attached to the emitter created instances of spheres that flow according to the spline. b) The same setup after activating the metaball modifier for both splines.

The following examples will show the power of procedural modeling combined with the node system in Houdini. We will only scratch the surface of this kind of modeling and spark your curiosity to experiment on this path.

Procedural Example 1

In this example, we will simulate the growing of an inorganic crystal. Although the effects are not based on algorithms of the crystal growth process, they can indeed create realistic results. We will create the nodes in Houdini, and it will take care of placing all of the required meshes/ volumes in the workspace. We cannot talk about steps since we are creating a graph based on nodes, but we will need to

a. Insert two nodes for the creation of primitives (a tetrahedron and an icosahedron).

b. Insert the node that will create scatter points on the surface of a base mesh where the smaller crystals will grow.

c. Insert the nodes that will duplicate smaller crystals in the place where our scattered points are.

In order to make the parameter change according to time, we will add the key $F in the corresponding text menu (one for each parameter we need to change). The $F means that the parameter will change according to FRAMES. No other instruction is necessary for this simulation.

Procedural Example 2

The first thing to notice is that we nearly copied the previous node system. This time we have two spheres instead of the platonic solid; we modified the rules of the parameters involved for one of the spheres, and we added a vdb remesh node similar to the metaball modifier used in Cinema4D. It creates a volume based on the copied spheres using an optimized volume (the vdb is tailored for work with volumes; it can be transformed later into a polymesh). The formula makes the sphere grow in time proportionally to the number of scattered points on its surface.

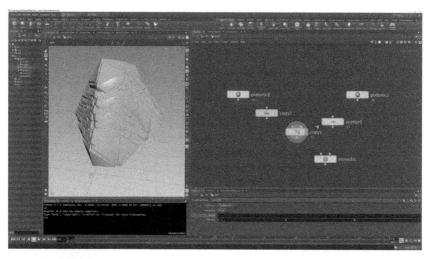

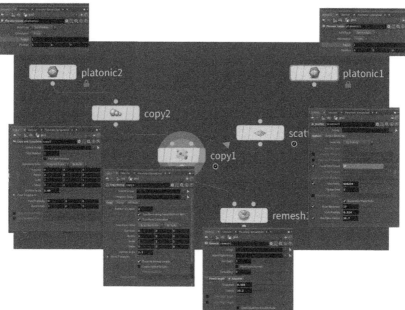

FIGURE 7.17 Top setup of the scene.

On the left, it is possible to see the result of the simulation after more than 100 frames. The six nodes on the top right take care of creating everything. Bottom setup of the scene. Parameters are used for each node in the simulation. You can notice that the only parameter that will change according to time is the total count of the scattered points on the surface of the base primitive ($F parameter). Also, a remesh node is linked to the copy node in order to create a cleaner final model.

FIGURE 7.18 Example of different step of the simulation.

A Few Final Thoughts

Just a few words about file formats. When you want to move animations from one 3D software to another, your first choice should be .abc, which tries to assure the maximum compatibility among software. If it is not possible to do that, I prefer to save each frame as a .dae file. If this option is not available, you can export all the frames in obj format as a last resort (you would need a lot of manual resetting to reproduce the animation that was originally created).

While creating simulation in Chimera, my choice is to use the dae format since it can save most features. If you use the obj format, export the scene in the Wavefront OBJ file format. Only surfaces are exported, not molecular models. A single color for each surface is exported to a separate file with the suffix ".mtl"; both displayed and undisplayed surfaces are exported.

Due to the wide range of tools available for creating animations, it is easy to get lost. Remember that the same principles apply as for creating still images:

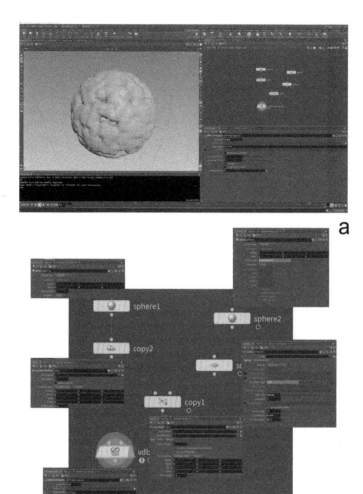

FIGURE 7.19 Top setup of the scene.

On the left, it is possible to see the result of the simulation after more than 400 frames. Again, the six nodes on the top right take care of creating everything. Bottom setup of the scene. Parameters used for each node in the simulation. You can see that we have two parameters that will change according to time (total number of scattered points and dimension of the base sphere). Here we have a vdb mesh remapper, which is similar to the metaball modifier in Cinema4D.

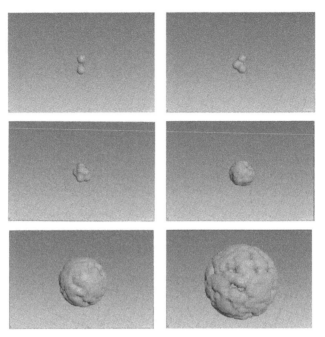

FIGURE 7.20   Example of a different step of the simulation.

1) As always, try to tell a story. Why do we need to use animation instead of a still frame? Focus on this point and highlight the main advantages of your choice.

2) Have a clear idea of the key frames you are going to create. Why do you consider this point of the animation more important than the others? Think about planning a small storyboard before beginning your animation. Your software will take care of creating the missing frames to animate the objects for you, but it does not know anything about the process involved in the animation.

3) Always split your animations in small intervals to maintain control of the process. It will be easier to see the strong and weak points of your animation.

4) Make use of colors, light, and depth of field to attract the attention of the viewer.

a. The brightest zone of your image will always attract the attention of the viewer.

b. Vivid and strong colors attract the viewer's eye.

c. Selective depth of field is always a good tool for focusing on the center of attention of your animation.

5) Choose your camera wisely. Wide angles are great for images that are more dynamic. Long lenses with shallow depth of field are great for simulating the use of a microscope

6) Keep in mind how your camera will move around the scene. You can adjust the details of the models in your scene according to the setup of your camera (i.e., if an object is not in focus you do not need to overdetail it, and you do not need to model what your camera will not see). It seems trivial, but it is easy to get lost while modeling a scene if you do not think about it.

7) The previous tip is valid also for shaders. You do not need to calculate extra detailed SSS on the surface of a molecule that is not the star of your animation.

8) Same tips for lighting. You do not need extra samples if your camera will not see them.

9) After calculating a simulation and rendering all the frames you need, reversing the order of the final composition can give the illusion of a reverse building of the process.

10) Plan time for the final compositing and try to be informed about what you will be able to change or not in post-production. The process for creating your final image should be iterative as we have previously seen. More specifically,

a. add the camera(s) to your scene,

b. add the objects,

c. add light(s) (lower samples),

d. animate everything you need (camera and/or lights and and/or objects),

e. create a previsualization animation,

f. make your change,

g. add shaders, and

h. retest the animation.

If everything is in good shape, go for a high-res render of your animation. Finally, keep in mind that displacement can change the shape of your objects, so add it before the final previz.

EXERCISES

Try to recreate the following animations:

- A simple turntable of a molecule (a virus or a DNA strand is perfect for this task). How can you change the settings of the camera and light to create different feelings for this small animation? Start changing one parameter at time (lights, shaders, etc.) and then choose a suitable camera. Try to apply different depth of fields.

- Create an animation involving many objects. How can you keep the viewer's attention? What techniques can you use?

- Recreate a docking animation. Simulate docking with your software of choice; import the object sequence in your 3D software and render it. How does it look? Is it too "flat"? Now focus on presenting each of the molecules alone using different camera movements; then follow with the camera docking. How does your animation look now? To keep going with this exercise, try to tell different stories of the same object sequence changing cameras and light settings.

- Choose one of your favorite animations in the field of scientific animation. Why does it work? Did the author tell a story? Which story is it? Can you recognize the topic we introduced in this chapter? Are the use of noise, expedients to keep the viewer's attention focused on the story, and the use of different speeds of animation related to the different process represented? Focus on what works and what you would have changed.

SUGGESTED READING

Allen, M. P., and D. J. Tildesley. 1993. *Computer Simulation in Chemical Physics.* Springer Netherlands.

Beane, Andy. 2012. *3D Animation Essentials.* Wiley.

Chopine, Ami. 2011. *3D Art Essentials: The Fundamentals of 3D Modeling, Texturing, and Animation.* Focal Press.

Elmore, Donald E. 2016. "Why Should Biochemistry Students Be Introduced to Molecular Dynamics Simulations-and How Can We Introduce Them?" *Biochemistry and Molecular Biology Education* 44 (2): 118–23. doi:10.1002/bmb.20943.

Jason, Sharpe, Charles J. Lumsden, and Nicholas Woolridge. 2008. *In Silico: 3D Animation and Simulation of Cell Biology with Maya and MEL.* Morgan Kaufmann/Elsevier.

Johnson, Graham T., and Samuel Hertig. 2014. "A Guide to the Visual Analysis and Communication of Biomolecular Structural Data." *Nature Reviews. Molecular Cell Biology* 15 (10). Nature Publishing Group: 690–98. doi:10.1038/nrm3874.

Jones, L. L., and R. M. Kelly. 2015. "Visualization: The Key to Understanding Chemistry Concepts." In *ACS Symposium Series* (Vol. 1208, pp. 121–40). https://doi.org/10.1021/bk-2015-1208.ch008.

Lasseter, John. 1987. "Principles of Traditional Animation Applied to 3D Computer Animation." *ACM SIGGRAPH Computer Graphics* 21 (4): 35–44. doi:10.1145/37402.37407.

Lok, Corie. 2011. "Biomedical Illustration: From Monsters to Molecules." *Nature* 477 (7364): 359–61. doi:10.1038/nj7364-359a.

Meindhardt, H. 2012. *The Algorithmic Beauty of Sea Shells. Uma Ética Para Quantos?* Vol. XXXIII. doi:10.1007/s13398-014-0173-7.2.

Miller, Katharine. 2011. "Visualization in Space and Time: Hollywood Style." *Biomedical Computation Review* (Fall): 9–11. Retrieved from www.biomedicalcomputationreview.org.

Milo, Ron and R. Phillips. 2015. *Cell Biology by the Numbers.* Garland Science Taylor and Francis Group.

Paquette, A. 2013. *An Introduction to Computer Graphics for Artists*. Springer. https://doi.org/10.1007/978-1-4471-5100-5

Prusinkiewicz, Przemyslaw, and James Hanan. 1989. *Lindenmayer Systems, Fractals, and Plants*. Springer-Verlag.

Wu, Dingbing, Aolei Yang, Lingling Zhu, and Chi Zhang. 2014. "Life System Modeling and Simulation." *International Conference on Life System Modeling and Simulation and International Conference on Intelligent Computing for Sustainable Energy and Environment.* Vol. 461. doi:10.1007/978-3-662-45283-7.

CHAPTER 8

Final Look

COMPOSITING

In the previous chapter, we highlighted what modern movies have in common with the first movies created using a zoetrope. We saw that although the media we use now are quite different, we need to create at least 24 frames per second in order to create a continuous image in the viewer's retina.

Digital frames are the analogous to film frames. What about render passes? You probably already know that before the ubiquitous use of computers, the word layers referred to physical cellulose acetate layers made of different materials stacked obtain the desired image. Think about adding titles to the cover photo of a magazine. How was it done? It was obtained by stacking (*in reality)* one acetate layer for the titles, one for the borders, etc. etc.

A similar process was performed for creating special effects on movies where different techniques relied on the layering of films that included the details needed for the final result. You can find more details (matte, rotoscoping, etc.) in the suggested literature at the end of the chapter. Today, we use a digital stack with lots of positive consequences for our work (just think how much faster the process is now).

In the chapter dedicated to rendering, we saw that while creating materials, we can focus on each individual phenomenon that happens when light interacts with the surface of an object. It is possible to create one image for each phenomenon, rendering a separate image for the diffuse, reflected, refracted light component, etc. If you then stack (add) these images together, you will reconstruct the final rendering image. In other words, using the additive properties of the behavior of light we can recreate the whole light interaction. Is it possible to record other information about our scene that can be used to reconstruct the final image? The answer is yes.

We are not limited to light behavior, but it is common for render engines to provide separate passes for the background, the Z-depth,

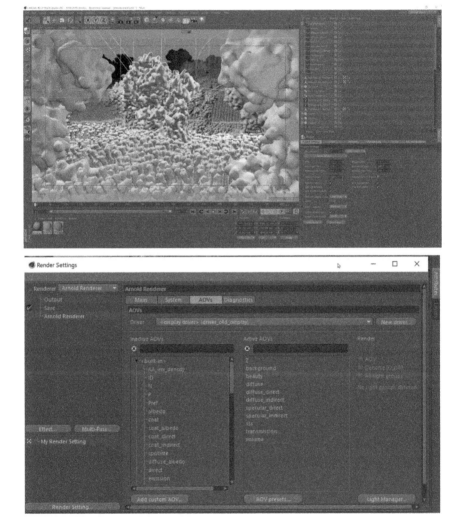

FIGURE 8.1   Setup of the membrane scene with all options offered by Arnold Render for creating render passes.

volume, shadow, effects (hair), ID for each object, and passes for NPR (toon pass, highlight, and rim light). In the Figure 8.1, we show all the options offered by Arnold Render.

In this chapter we will focus on the composite of static (single frame) images, but all principles can be applied while compositing movies where the arsenal of post enhancement and effects that can be applied is even wider due to the added dimension of time.

The first thing we can do is recreate what is called a beauty image starting from its component. We do that to have more control (among other reasons). After recreating the image, we would probably would like to polish it by adding vibrancy, checking color saturation, changing levels, adding bloom, adding glare, adding fog that we could not create during the render, adding floating particles to our animation, etc. Instructions for performing these actions will be presented in the second part of this chapter.

COMPOSITING PART 1

So why we should bother compositing an image when our software can produce a beauty image? The answer, as you should already know is that it depends. Sometimes we are fine making a few adjustments to the beauty pass; sometimes we need to have total control of each component of the image and need to use a more traditional workflow. Recreating a beauty part from its components lets us be more flexible and eventually make minor adjustments to the image without rerendering everything. One of the most common examples is rendering a Z-depth pass to change the depth of field (DOF) of your image in post or to correct a washed look. Even if the renderer breaks the wall of real-time representation, it can be good to know that a quick fix is possible. You can always render an occlusion pass and apply it in post, adding (we will see what we mean by adding) it as a new layer in your favorite composite software.

Before going on, it is necessary to introduce the tools we use. While compositing images we can choose two paths. One is using a layering system similar to the one included in Adobe Photoshop/Adobe After Effects; the other option is using a node system like the one used by Nuke/Fusion.

Each one has its own strong points, and I love using them both. For animations, I prefer to use the Nuke/Fusion approach, and for quick adjustments, I use Affinity photo.

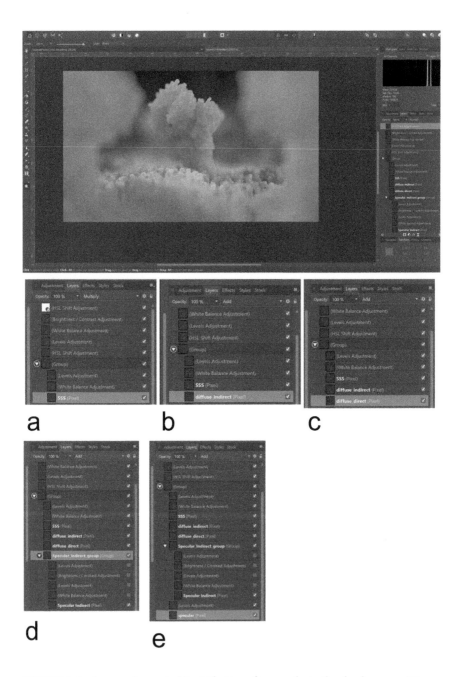

FIGURE 8.2　Images imported in Affinity software photo for final compositing.

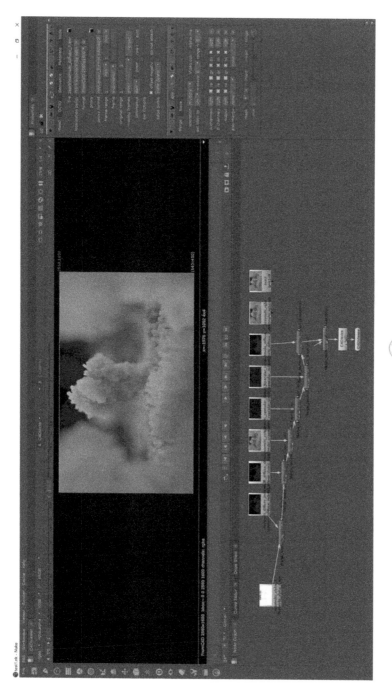

FIGURE 8.3 Images imported in Nuke for final compositing.

Example 1 Compositing of an Image

We now rerender the project of creating an artistic illustration of a membrane. We set the rendering buffer as shown in Figure 8.1 and use a common set of additive Arbitrary Output Variables (AOV) that consists of:

Diffuse_direct

Diffuse_indirect

Specular_direct

Specular_indirect

Coat

Transmission

SSS

Volume

Emission

Background

For the image in the example, we will not use them all and can exclude some of them from the calculation. How do we know which pass to exclude and why? Keep in mind that using AOVs is useful for the reasons already presented. When your computer is going to render using passes, it will use more memory than when it is rendering only the final beauty pass. Adding extra layers that are not present/used (and because of that will render in black) is a waste of memory (i.e., if there are no SSS materials it is pointless to add an SSS pass). Now it is time to reconstruct the beauty pass using the layers rendered.

If we would prefer to use a layering system we can setup all the layers to 100 percent opacity and add mode in Affinity photo (other software may have different settings, sometimes called linear add) while the Z-depth pass should be set to multiply.

While working with a node system we will use the same approach, adding all the passes together. Cinema4D lets us save time while

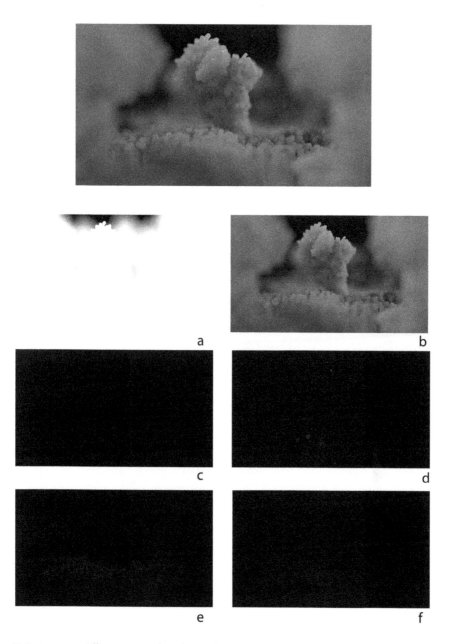

a

b

c

d

e

f

FIGURE 8.4 All passes rendered. Final Image composed by the following passes
a) Z-depth. b) SSS. c) Diffuse_direct, d) Diffuse_indirect e) Spec-
ular_indirect f) Specular_direct.

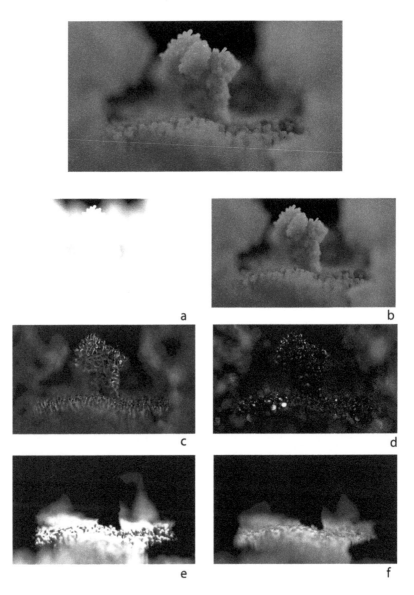

FIGURE 8.5 Same as Figure 8.4, but in order to highlight the individual
contribution of each AOV they are presented using increased
intensity.

rendering in passes, having already created the comp file, which is ready to have after effects, motion, nuke, or digital fusion imported.

After importing all the passes to Affinity designer, we will probably notice a pixelated look in some of the passes. This will depend on what render engine is used and how it internally manages the rendering process. In order to avoid issues, it is useful to apply a blur filter before creating the final comp. Also, we can adjust the brightness and contrast, levels, colors hue, and saturation of the different passes in order to achieve the desired look.

At this point, we should see no difference between a previous rendered beauty pass and the image obtained by merging our layers. Before introducing FX passes, we will show two other examples. In the first, we add an occlusion pass. An occlusion pass is obtained using a neutral light grey shader to the image using only a skydome light. The image obtained can be composed with the beauty image to add contrast and eventually correct the presence of unrealistic white areas that sometimes can be created as an artifact by global illumination.

It is interesting to notice that in order to create the zpass we switched to the intern physical render and applied a simple neutral grey shader. You can combine the pass made in different renders and use this feature to your advantage.

FIGURE 8.6 "Raw" composite before performing any adjustment.

FIGURE 8.7 Final composited image after adjusting levels and contrast for all the passes in comparison with beauty pass obtained by means of Affinity designer.

FIGURE 8.8 Final composited image created using nuke.

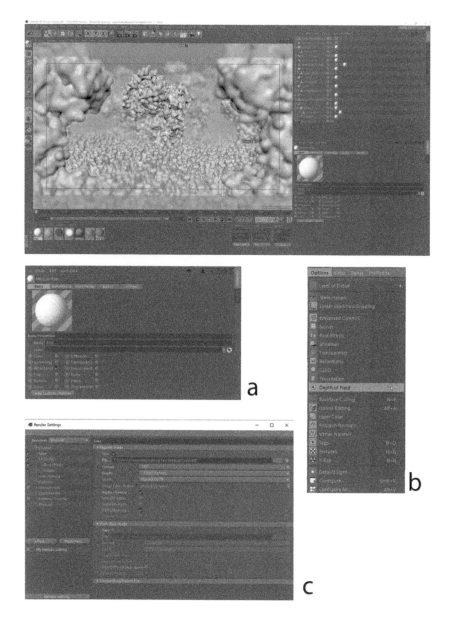

FIGURE 8.9 Settings for rendering the occlusion pass. Neutral shader applied to all of the objects in the scene. Occlusion pass added to beauty pass. Top beauty pass, middle occlusion pass, bottom, composite. The effect is overdone for the sake of clarity.

FIGURE 8.9 (continued).

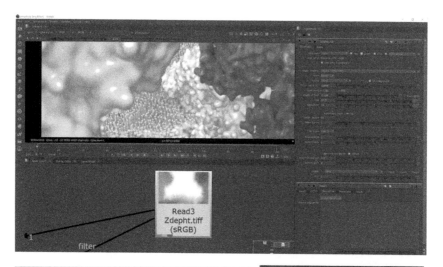

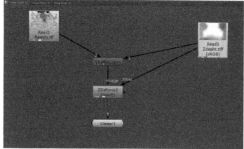

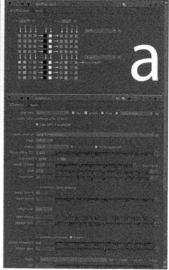

FIGURE 8.10   Defocusing added using the software nuke.

The same effect can be obtained via the layering technique, but the reader should carefully check if the software is able to perform a depth blur action. In order to better enhance the effect of the zfocus, the beauty pass uses different colors for one selected protein while the other molecules are represented using a neutral shader. For compositing the image in nuke, we used a shuffle node set as seen in a) and a ZDefocus node using the settings shown in b).

Another easy and useful correction done in post is the defocusing of an image via z-pass. We can use the Z-depth pass previously obtained in order to composite a new image sparing the time to recalculate it while rendering the new image.

COMPOSITING PART 2

After creating our beauty pass, we are ready to add FX and change the overall look according to our personal taste. To explain how to proceed we will add a glow filter to the specular_indirect pass in Affinity designer and repeat an analogous process to the same pass in Foundry nuke to get a "sparkly" look.

FIGURE 8.11   Image created after applying a glow filter to the specular_indirect aov in Affinity designer.

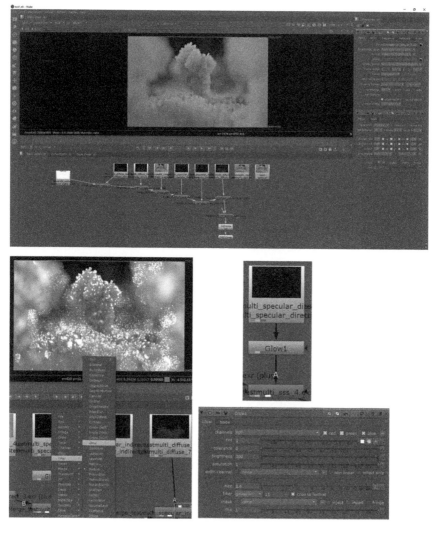

FIGURE 8.12   Image created after applying a glow filter to the specular_indirect AOV in Foundry nuke.

Instead of modifying the rendered channel, we can add new layers to fake volume light, fog, particles, etc. in order to obtain the desired artistic representation.

EXAMPLE OF NPR RENDERING

NPR is very common for working with biomedical illustrations to represent concepts using a non-realistic render approach. The first pioneers of biological/chemistry artistic illustrations painted their representation by hand (one of my favorite books is Pauli's *Architecture of Molecules*. We can still admire the beautiful nanomachinery created by David Goodsell using watercolors. These representations are beautiful and indeed have a few strong points that make them useful today. It is easy to make the viewer focus on the molecule of our choice. The use of silhouette lets us focus on the shape of the molecules. The level of detail can be varied ad hoc for each molecule in the representation.

In the next example, we will see how we can combine the use of 3D and 2D software to use this style of representation. The good news is that we do not need to paint everything from scratch (tracing all the molecules) and that we can use the model already imported into our 3D software. Many render engines offer NPR solutions, including Solidangle Arnold. The steps for creating the NPR are the following:

1) Open your "realistic" scene.

2) Select all shaders.

3) Delete their assignment to the object in the scene.

4) Create two NPR shaders (in Arnold toon shader refers to the NPR shader) with a very strong base color (i.e., one red and one blue) and apply them to our object of interest.

5) Add a dome light to the scene (plugins, Arnold, light, skydome) to get rid of the previous lighting assets.

6) Go to options rendering and add trace contour in the filter menu.

You can start the Arnold real-time preview in order to inspect how the image will render with these default settings. We now adjust the specularity of the material in order to enhance the visibility of the features of the molecules that we would like to represent.

The image is now ready to be rendered in high resolution for the desired effect using a 2D paint software. My personal choice is to select

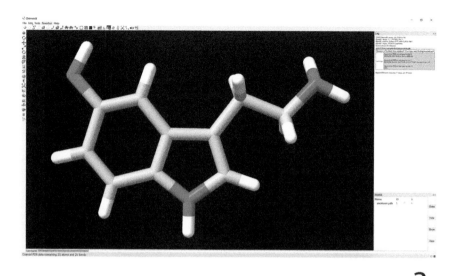

FIGURE 8.13 Toon material used in order to obtain the silhouette of the molecules that will be finalized via 2D painting software.

FIGURE 8.14   Final NPR after adding highlights manually using pastels.

a suitable (small) palette before coloring the image and then apply watercolors and/or colored pencils. I generally avoid oils and airbrushes or wide brushes to make sure that I do not obtain a 3D look that will make the whole process pointless. Notice that the scientific software previously presented has render capabilities that are very helpful if we want to create an NPR rendering. In the following example, I will recreate a pastel painting of a molecule of serotonin.

First, we change the setup for visualization as shown in Figure 8.14, applying soft light and getting rid of the silhouette.

The rendered image is then imported into the software Rebelle, composited with a paper background.

TIPS

A few words about color spaces and linear and non-linear space: You can find many threads in many different media (web forums, discord, slack channels, etc.) about the correct workflow for dealing with composite passes and rendering in general. We can start with the idea that our software will record the information about each pixel rendered. This information can be stored in different bitmap file formats that record the red, green, and blue lights and the transparency of each pixel. We need to know that our eyes have different sensitivity to these colors, so after creating our render we can assign a color space in order to "cook" the colors of the image and give a more realistic depiction. We cannot perform this process for each pass *before* compositing it; otherwise, we will introduce inconsistencies into the final render due to the way our composite software adds the images together.

In summary, when we are dealing with compositing, our images need to be generated in every step (this includes the texture we use for creating the shaders of our material) using a linear workspace. As a final note, I will describe the minimum post-processing I perform on the images I render:

1) Import all the passes in my node editor.

2) Check each pass and adjust the levels in order to get rid of washed colors.

3) Check the vibrancy and saturation of each pass (layer).

4) Check the corners and dark areas of the image for artifacts.

5) Correct for total black or total white in the image to avoid a CG look.

6) Check the color space used, and apply a specific color profile depending on the final media where your product will be published.

EXERCISES

- Try to rerender one of the scenes you already created, splitting it into AOVs. Composite the layers using both workflows reviewed in this chapter: classic layer approach and node-based.

- Create a template for your own composite project in order to find the approach that better suits you.

- Create a simple test scene for your shader. Try to render different material in order to better understand the components of light involved while using different materials.

After getting better at recreating your beauty passes, try to add fx passes. Start with a fake volume light. What is the quality of the final composite? Does it look more or less "artificial"? Now try adding a grain noise and compare this to the original beauty pass. Is the quality of the final image improved?

SUGGESTED READING

Birn, Jeremy 2015. "Lighting and Rendering." In *Lighting and Rendering*. New Raiders.

Brinkmann, Ron. 2008. *The Art and Science of Digital Compositing: Techniques for Visual Effects, Animation and Motion Graphics*. Morgan Kaufmann Publishers/Elsevier.

Lanier, Lee. 2013. *Digital Compositing with Nuke*. Focal Press.

Lanier, Lee. 2015. *Compositing Visual Effects in after Effects - Essential Techniques*.

Menache, Alberto (2011). "The Motion Capture Session." *Understanding Motion Capture for Computer Animation*, (2), 75–134. https://doi.org/10.1016/B978-0-12-381496-8.00003-2.

Okun, Jeffrey A., and Susan Zwerman. 2010. *The Ves Handbook of Visual Effects: Industry Standard Vfx Practices and Procedures*. Taylor and Francis.

Wright, Steve. 2013. "Digital Compositing for Film and Video." In Kerlow, Isaac Victor, *The Art of 3D Computer Animation and Effects*. John Wiley.

CHAPTER 9

Professional Practices

T he world of biochemical illustration is quite vast, and it would be difficult to present all of its possible careers. In this chapter, I will follow the same approach I used in the introduction. I will write about my personal experience as a scientist, an illustrator, and a teacher.

As a scientist, I understood the nature of the illustrations I see in my everyday job. It is quite easy to take for granted that the representations we see are photographs of the molecules we work with. It is easy to forget that we are not dealing with macroscopic objects and that they do not follow Newtonian physics. It is also easy to forget that we are looking at maps that represent shapes in a fixed moment in time while, as all scientists know, molecules follow (until this is disproved) the indetermination principle. This gave me the chance to rethink the models we use. They are tools; they are not fixed in time; they can and must change and are part of the reality we create for getting a better understanding of nature. Recreating images gathered with different techniques gave me the opportunity to learn as much details as possible about them. As an example, to reconstruct a setup for macro-images, I needed to know at least the schematics of a microscope. Where does light come from? Is it direct? Are there surfaces that can diffuse light? Is light polarized? Are there any render engines that can recreate "complex" light interaction with matter?

I followed the same process while setting up a shader for the Secondary Electronic Microscope image microscope (SEM). How are the SEM images obtained? What kind of light interacts with matter in this case? Where are the lenses in an electronic microscope? (They are electromagnetic lenses and have fixed positions.) How can we define the "diffuse" color of an object under SEM (the Fresnel effect we see)? How does an atomic microscope work? Can we reproduce "topographies" in 3D (with all the subsequent techniques we can use for creating displacement maps, adding displacement and bumps, etc.)?

A very important task shared by illustrators and scientists concerns finding patterns and connections in their work. This refers to images, textures, and ideas while linking distant subjects (i.e., the same displacement technique we used for creating a terrain is helpful for recreating an electronic microscope image). Growing mechanisms are ubiquitous in nature. If we simulate a cellular *growing mechanism*, we can use the same algorithms to simulate any growing process. Clustering is another example. How do particles *flock* together? How can we simulate this? All of these challenges stimulate finding answers in different fields of study; keeping an open mind is always positive for a scientist.

Finally, I need to use 3D spatial abilities all the time, in 3D, chemistry, and CG, using different software. The need for seeing the world upside down is quite often very helpful in order not to get stuck in a specific hypothesis.

Let us now see how being an illustrator can benefit you as a scientist. Apart from the obvious skill of designing the illustration for your papers and understanding the art guidelines any researcher must follow while submitting a paper, it helps clarify the process of disseminating your results (e.g., it would give you headache to explain to colleagues that using a 1200 dpi image cannot be obtained resampling a 72 dpi jpg downloaded from a website).

We all know that while being buried in our lab we often neglect presenting ideas in papers, thinking that everyone already knows what we know. Illustrating concepts makes you reflect on the basic questions you should ask yourself while writing a paper: What is my research about? What results have I obtained from my research? Who can benefit from them? How can I explain with clarity my results?

Illustrating concepts from scientific domains different from your own should help you exchange ideas with other scientists, keep your attitude humble, and force you to get as many references as possible to maintain high quality. Illustrating your own concepts will also let you experiment with your abstract abilities and occasionally step out of your comfort zone.

Finally, I would like to report the answers to the most common questions asked by my students:

Q. *What software do I need to use?*
A. As I reported in the beginning of the book, you should change your question according to what you would like to represent and how. There is no software with a magic button; just learn skills with different tools to better depict your ideas. It is also important to understand the basic paradigms of each software in order not to be limited by your tools (if you only have a hammer, everything will look like nails!).

Q. *Do I need to be proficient in hand drawing to use 3D software?*
A. The more techniques you know, the better, but it is not strictly mandatory to know hand drawing.

I do encourage my students to take traditional painting courses in order to understand how to *observe* and not to merely *see* whatever helps you get the desired result (and now violating ethical principles). It is said that the old masters used *camera obscura*, but this fact does not demean their value at all.

Q. *Is it easy to create an animation after I set up and render my still image scene?*
A. It depends. If you are looking for a non-realistic docking animation or a flythrough animation, it can be easier than you think, but creating animations of more complex behaviors would probably need rethinking of the modeling process of your scene.

Q. *I have a few ideas on my scene, but where can I get references?*
A. One of the most common mistakes my students make is to gather insufficient references for their projects. Gathering references is one of the most important parts of a project and it should not be undervalued

If you are illustrating a process involving a molecule, you should try to find all the recent papers on it that will remand you to the best reference on the topic. Also, my order of preference about scientific literature is to first search in books/libraries (electronic or brick and mortar), specific scientific databases, and as a last resource general search engines.

Q. How do I model xy?
A. Did you research before starting modeling? This is generally the first answer I give. What do we know about xy? Is the product of a well-defined process? Is it "static" in time, or will it change? Do you need to animate it, or do you need only a still image? Are you sure you need to model xy, or you can find it in a database? If you have the answers to these questions and only need a still, you can choose among nurbs modeling, polygonal modeling, and procedural modeling.

Q. how do I set up lights?
A. Think of your 3D scene as a photographic studio or movie set (or even a painter's studio). You should already know the questions I'm going to ask: Have you gathered references for your work? This is not limited to photos or movies (or storyboards or painting) with the same subject as yours (that would be not useful since we hope you are not going to replicate a work that already exists) but refers to media you like for their photography (staging). While looking at a picture/movie that you like, ask yourself where the photographer/camera is. Where are the lights? How is the subject lit? Most of all: What emotions do you feel when you look at it?

Q. I don't like my skill on shaders. Can you help me?
A. Remember that the final appearance of an object is due both to the lights and materials you used in the scene. Therefore, if you are looking to recreate a special look/feeling, you should study both. Generally, you start setting up the lights and then set up the materials. If you have "references" for the appearance of the material you like, try for a photorealistic reproduction of its characteristics; otherwise, use a shader that is a function of what you would like your audience to "feel."

Q. Where may I find freelance job offers?
A. Do you have a portfolio? If the answer is yes, and you already know what kind of work/illustration you would like to create, you can try

contacting specific editors, industries, or institutions (quite numerous in the field of science/biomedical). As always, networking works best.

Q. What FX should I use in post to make my image/movie stunning?
A. Again, there is no magic button, so it depends. Try to avoid cliché ones. Remember that your image should pass the test of time. The rule is not to overdo. Sometimes we forget that we do not have only one alterative; you can prepare different takes to test your results, and if your client is asking for something special that he/she cannot define, ask for examples and references.

Instead of using too much FX remember to

- create images with good contrast and eventually change them in post,

- always check the composition of your image and crop it if needed,

- always check the levels of your images and try to avoid showing pure black or white (that are quite unusual in the "real" world and give to the viewer an immediate feeling of an "artificial image").

- adding a bit of noise is ok since in life even in the cleanest laboratory there is always a bit of dirt/noise; our images should reflect that, but do not spend time adding too much noise to a perfectly cleaner render! If this happens, lower your render settings and noise will come in by itself!

- the same goes for chromatic aberration, lens flare, vignetting, etc. Photographers spend a lot of money buying the best lens and avoiding these defects, so adding a few for realism is ok, but don't overdo.

A FEW REMARKS ON SKILLS

Try to improve your skills, and never forget that the software you use is only a tool. Try to test new software during your working year, and do not rely on one specific software. It is useful to know every detail of your workhorse, but do not be afraid of experimenting. Use and sharpen your skills for creating a personal project. Set yourself small goals, and remake older projects using new tools. Always try to understand the concepts and reasons behind the introduction of new tools. What are their stronger points? What can this new software do that was not possible using your workhorse? Think about procedural modeling. Is it worth it? Who uses it? Why was it introduced? Can I model thousands of objects individually?

Create a safe environment where you can experiment with your ideas. Try to be part of a community where you can exchange ideas and try new challenges. This will help you grow. It does not need to be overly competitive, rather a test of your skills.

Always try to be nice and create a network of contacts. This should be never underestimated. Remember that the industry prefers a less skilled artist who can work in groups to a "genius" loner. Also, if you do not have the time to take all the work offered to you, and if you know someone that can do it, you can suggest his/her name to your client. It is a win/win for both of you.

About portfolio/social/forum/websites: Keep in mind that you will be judged by your weakest work and that is easy to be tempted to show only your latest work. If you have good old projects to show, do not be afraid to post them.

Be consistent in the work you show, and try to develop a personal style. It is ok to be a generalist, and it is a good skill to have several examples of projects; it is also important to show how you can make your contribution unique in the work you take. I prefer to have a personal website and to show my work on selected CG websites. Keep your website minimal in order to update it quickly when you need it, and do keep a balance between the time you spend on social platforms and how much you learn/grow. Time is a limited resource, and you should think about where it is worth spending it.

Keep a balance between personal life and work. It is easy to forget to create long-lasting links when you are young. You will never regret time spent on better relations with your relatives and the people you love.

A personal note on your "hardware": your body and health. As a scientist or illustrator, you can have the chance to have lot of freedom about how you schedule your work. This sometime can push you through crunch time. The first obvious remedy is to try to schedule your work in order to avoid it, but if you cannot, take precautions to stay healthy. It can be crazy, but if you visit online artist communities and talk with artists "in person," eventually you will talk about health issues.

In a few words, try to keep fit! Do not overestimate the importance of a proper diet, necessary sleep, and the health of your backbones, wrists, and eyes. Try to prevent eyestrain; try to walk whenever you can, and if you feel pain to elbow and wrist try to find the most suitable controller that works for you (due to a problem with my wrists, I needed to switch my mouse for a trackball; I will never go back. Also, in my everyday work as scientist, I always use two large displays. The ratio between their cost and benefit is positive for sure!

One last word about your daily attitude: Try to keep fresh eyes on problems and to nurture the attitude for seeing connections between the scientific concepts you illustrate and everything else in your everyday life. Better ideas for illustrations come when you let your mind wander. Happy drawing!

Index

Page numbers in *italic* indicate figures.
Page numbers in **bold** indicate tables.

3D elements of a model, 56–57
3D matrix, 60
3D modeling, key concepts, 44
3D spatial abilities, 204

A

Abstract models, 21–22
Abstraction levels, 75
Academic publications, art
      guidelines, 204
Affinity designer, 191, *192*, 196,
      196, 197
Affinity photo, 185, *186*, 188
Alanine file, 44
Animated model, exporting, 5
Animation, 149–182
    artistic aspect, 152
    autokeying options, 158, *159*
    camera choice, 179
    conclusions/recommendations,
      176–180
    core concepts, 155
    DNA assembly example, 169, *170*
    file formats, 176
    frames per second, 149, 183
    inorganic cell self-assembly, 157–158,
      159, 160
    key frames, 149, 157–160
    living cell example, 162–165
    mesoscopic scale, 156

    morphing molecule, 150–157, *151*, *153*,
      154–155, *156*
    non-linear variation controls, *164*
    non-realistic simulation, 156–157
    particle simulations, *165*, *166*,
      167, *168*
    particle trajectory tracing, 169–171,
      172, *173*
    plant growing process, 171
    principles of, 149–157
    procedural modeling, 171–174, *175*,
      176, *177*, *178*
    software menu options, *151*
    task difficulty, 157, 205
    turntable of DNA fragment, 157,
      158–160, *161*, *162*, *163*
    use of noise, 160–161
Anti-aliasing, 143
Arbitrary Output Variables (AOV), 188,
      190, 197
Arnold Render, *184*, 185, 198
Artistic interpretations, animation, 150
Atomic microscope, 113–114, 204
Atoms
    bonding and energy, 17–20
    color associated with elements, *14*, 15
    and molecules, 30–32
    smallest part of models, 15
    as spheres, 20, 29
    as transparent clouds, 29
    volumes of, 10

Audience
  communicating to, 13
  consideration of, 110, 113
  and visual appeal, 14
  visualization tools, *14*
Augmented reality, 10–11
Autodesk 3D Studio Max, 4
Autodesk Fusion (360), 70
Autokeying options, 158, *159*
Avogadro, 32, *34–35*

B

Ball and stick models, 10, 15, 20, *21*,
  33–36, 44
Beauty image, 185, 188
Bevel shaves, 63
Bidirectional reflectance distribution
  (BRDF), 142
Binding partners, 22
Biology, water as ubiquitous solvent, 88
Biomedical Illustration Workshop, 1–2
Biomedical illustrations, 198
Biomolecules, 20, 22
BioNumbers database, 76, 157
Bitmap, 92
Black and white images, 92, 97, 207
Blobby surfaces, 22
Blood cells, 122–124
Blood clot, 123, *125*
Blur filter, 191
Bond-centric models, 20
Boolean options, 65, 70
Box modeling, 57–61, 116, *117*
BPQS layer, 116, *117*
BRDF (Bidirectional reflectance
  distribution), 142
Bump maps, 139–141

C

CAD files, 70
Calibrated workflow, 93
Calibration (color), monitor, 93
Camera *see* Virtual camera

Carbon, 20
Carbon nanotube, 38–39
Cartesian coordinate system, 30
Cartesian space, 46–48
Cartoon representation, 22, *24*
Cathedral lights, 82
Catmull-Clark subdivision
  methodology, 62
Caustics, 113
Cave paintings, hominids, 27
Cell biology, and Molecular biology, 11
Cell membrane, artistic interpretation of,
  128–134, *144–145*
Cellimagelibrary.org, 122
CGChallenge, HIV virus visualization, 1–2
CGMA (Computer Graphics Academy), 4
CGSociety, Biomedical Illustration
  Workshop, 1–2
CGtalk forum, 1
Chains, 22
Chemical concepts, representation,
  multimedia content, 10–11
Chemical language, 9, 15
Chemistry
  representations in, 9–11
  representing in 3D space, 30–32
Chemoinformatics (Gasteiger and Engel), 30
Chimera, 5, *17*
  command line console, 152
  DNA representation, *20, 21*
  graphic user interface, *40*
  high surface representation, *135*
  importing models, 139
  importing PDB files, 40–43, *43*, 128
  morphing molecule animation,
    150–157
  organic molecules, 33
  scripts, 150–153
  stop-motion animation, 150
  virus representation, 40–42, *41*
Chromatic aberrations, 113, 207
Cinema4D, 4, 5
  animation tools, *164*
  box modeling, selection options, *61*
  color selection tools, *86–87, 89, 90*

graphic user interface, *46*
light menu, *81*
modeling actions, *66*
modifiers, 65
non-linear variation controls, *164*
particle trajectory tracing, 171
primitives, *58*
rendering passes, 188–191
scale of scene, *77*
scripts, 71
turntable of DNA fragment, *161*
virtual camera, *48*, *51–52*
Clichés, overdone in images, 146, 207
Clusters, 22, 204
Collapsing, 64
Color blindness, 91, *92*
Color space, 201
Colors
    associated with elements, *14*, 15
    calibration, 93
    Corey-Pauling-Koltum (CPK models),
        conventions, 10, **16–17**, *17*,
        24, 97
    deuteranomaly, 91, *92*
    diseases, 94
    free choice, 95
    mood of scene, 85–87, 88, 94–95
    palette selection, 88, 94
    primary, 94
    printed media, 92, 93
    random hue modifier, 128
    realistic depiction, 201
    scene setup, 88–95 tools to modify,
        86–87, *89*, *90*
    and shaders, 110
    tissues and blood, 94
Compositing
    beauty image, 185, 188
    behavior of light, 183–184
    digital stack, 183
    example image, 188–196
    glow filter, 196, *196*, *197*
    node system, 185, 188
    physical cellulose layers, 183
    rendering passes, 184–185

    software layering systems, 185, 188
    your image, 93, 179, 183–196
Concept visualization, 9
Console, 71, 152
Constant shading, 58
Convolution surface models (CS), 21,
    22, *23*
Coordinates system, creating, 46–55
Copper crystal shape, *19*
Corey-Pauling-Koltum (CPK models), 10,
    16–17, *17*, *24*, 97
Covalent bonds, *18*, 20
CPK models (Corey-Pauling-Koltum), 10,
    16–17, *17*, *24*, 97
Crystallography, 15, 33, 36, 39–42
CS (convolution surface) models, 21, 22, *23*
Cubes, creating, 57–61
Curves, and modifiers, 63
Cutting, 62
Cylinders, *17*, 19, 20–21, 29

D

Dalton systematic series, 9, 15
Data
    calculations, 70
    scale range, 11
    and scripts, 72
    visualization, 9, 13
Decimation, 63
Default lighting, 29, 79
Depth of field (DOF), *52–53*, 185, 191,
    195, 196
Deuteranomaly, 91, *92*
Diffuse channel, 110, *112*, 114, *115*,
    116, 138
Digital images, pointillism, 91
Direct input, 61–68
Displacement, object shape, 180
Displacement maps, 139, *140*
DNA replication (human body), 157
DNA representation
    assembly animation, 169, *170*
    chains, 44
    Chimera, *20*, *21*

surface representation, *21*
turntable animation, 157, 158–160, *161,*
     162, 163
DOF (depth of field), *52–53*, 185, 191,
     195, 196
Dots per inch (DPI), 91, 92
Downloads
   3D structures, 31
   additional files from author, 8
DPI (dots per inch), 91, 92

E

Edges, 57, 63–64
Educational resources, 13–14
Edutainment world, 3
Electronic microscope images, 97,
     113–114, 122–128
Elements, bonding and energy, 17–20
EMBL (European Molecular Biology
     Laboratory), 33, 42–43
'Enlarging', 11
Environments, 113
EPMV, 31, 44–46, *45*
European Molecular Biology Laboratory
     (EMBL), 33, 42–43
Extrusion, 62, 63
Eyestrain, 209

F

Faces (object), 57, 134
File formats
   3D objects, 29, 30, **30**
   animation, 176
   CAD, 70
   export options, 33
   PDB, 34–35, 36, 40–42
   printed media, 93
   uncompressed image, 6, 92, 200–201
   vectorial, 92–93
File management, 6–8
File trees, 7
Freelance jobs, 206–207
Fresnel gradient, 113–114, 128

Fusion, 185
FX passes, 191

G

Gabedit, *18*
Gaussian Surface model, *23*
Global illumination, 81
Glow filter, 196, *196, 197*
God lights, 82
Graphic user interface, 71

H

Halos, 18
Hand drawing, 205
Hand painted work, 198, 200
Hardware, 8, 93, 209
Houdini, 4, *58*
   animation tools, *164*
   box modeling, selection options, *61*
   graphic user interface, *47*
   importing models for animation,
        153, *156*
   light menu, *81*
   modeling actions, *67*
   node system, 174
   non-linear variation controls, *164*
   and procedural modeling, 174
   scale of scene, *78*
   and scripts, 72
   special effects, 5
   virtual camera, *48*
Hue, saturation, value (HSV), 88
Human anatomy, knowledge of, 11
Hydration surface representation, *135,*
     138, 138
Hyperballs representation (Chaven), 21

I

Illustrations
   for diagrams, 113, *115*
   scientist's understanding, 203
   your own concepts, 204–205

Image resolution, 91–92
Inorganic bonds, 20
Inorganic cell, self-assembly animation, 157–158, *159*, *160*
Inorganic macromolecules, 32
Inorganic molecules, 32–33
Inorganic structures, drawing, 36–38
Ionic bond, 19

J

Journal covers, 1, 183

K

Kelvin temperature, 87, 88
Key frames, 149

L

Layering systems
    physical cellulose layers, 183
    software, 185, 188
Learning paths, 3
Lens flare, fake, 146
Ligand excluded surfaces (LES), 21, 22
Light decay, 85, *85*
Light wavelengths, 15, 97
Lighting, 29–73
    customization menus, 82–83
    default, 29, 79
    example menus, *81*
    face orientation, 57
    fake reflected, 142
    flipped normals, *112*
    interaction with materials, 95–101, *98*
    occlusion, 142
    path from sun to retina, 97–98
    post-production, 82
    ray depths, 134
    real life, 82, 85
    reflections, *98*
    setting up, 79–88, 206
    sources of, *83*
    surface scattering, *100–101*

and translucency effects, 113
    virtual camera, *131*
    workflow, 5–6, 114–121, *121*
    wrong intensity of, 76
Liquid environment, 87–88, 128, *166*
Lofting, 63

M

Magazine covers, 1, 183
Material descriptions, 98–99, 113–114
Maya, 31, 68, *68*
Media selection, scene setup, 91–92
Medical field, imaging techniques, 11
Mesh (object), 57
    modification of faces, 62–63, *64*
    modifiers (whole mesh), 65
    remapper, *177*
    transformation matrix, 60
Mesoscopic scale, additional tools, 23–26
Metaballs, 22, 29–30, 123, *165*
Metallic bond, 18
Metaphors
    artistic interpretations, *26*
    molecular visualization, 15
    overview, 9
    visual, 13, 110
Metaprogramming languages, 70–71
Methane, PDB file, 30–31
Microscope schematics, 203
Modifiers
    and animation, 157
    and curves, 63
    noise generation, 160–161
    random, *165*
    vertex, 63–64, *65–66*
    whole mesh, 65
Modo, 4, 5
    animation tools, *164*
    box modeling, selection options, *61*
    diffuse texture map, *136–137*
    graphic user interface, *47*
    importing PDB files, 138
    light menu, *81*
    modeling actions, *66*

primitives, 58
scale of scene, 77
and scripts, 72
turntable of DNA fragment, 162
viewport setting, 59
virtual camera, 48, 49–50
Molecular architecture, 22
Molecular biology, 11
Molecular map, 154–155
Molecular modeling software, 150
Molecule bonds, 10
Molecule coordinates, 5
Molecule skin surface (MSS), 21, 22
Molecules
   animation, 150–157, 151, 153,
      154–155, 156
   and atoms, 30–32
   NaCl representations, 19, 110, 111, 158
Mood, and colors, 85–87, 88, 94–95
Movies, compositing, layers, 183
MSS (molecule skin surface), 21, 22
Multimedia content, 10–11, 13–14

N

NaCl molecule representations, 19, 110,
   111, 158
Nanoengineer Nanorex, 5, 33, 38–39
Nanomachinery
   definition, 32
   how to draw, 32–39
Networking, professional, 208
Newtonian physics, 150, 203
N-gons, 68
Node system, 174, 185, 188
Noise generation
   and animation, 160–161
   anti-aliasing, 143
   displacement maps, 128, 139, 140
   too much, 207
Noise maps, 128, 139, 140
Non-linear variation controls, 164
Non-realistic simulation, 156–157
non-uniform rational Bezier spline
   (NURBS), 44, 68–70, 68

Normal maps, 139
Normals, 57, 69, 112
NPR rendering, 198–200, 199, 200
Nuke, 185, 187, 192, 195, 197
NURBS (non-uniform rational Bezier
   spline), 44, 68–70, 68

O

Object size, blurring the building
   bricks, 30
Objects
   exterior boundary, 61
   material description, 98–99, 113–114
   number of, 75
   scale in representation, 75–78, 76, 77,
      86–87
Occlusion, 142
Occlusion pass, 185, 191, 193–194
Occlusion surface, 24
Online artist communities, 209
Organic chemistry, 20
Organic macromolecules, 32
Organic molecules
   drawing, 33–36, 34–35
   intended analysis task, 20
Origin, Cartesian space, 46–48

P

Palette selection, 88
Paper and pencil, 5, 6, 27
Parsers, 31–32
Part modeling, 61–62
Particle clusters, 22, 204
Particle emitters, 157, 167, 168, 172, 173
Particle simulations, 165, 166, 167, 168
Particle trajectory tracing, animation,
   169–171, 172, 173
Pastel painting, serotonin molecule, 199,
   200, 200
PDB (protein data bank), 5
   education posters, 128
   entry 1yti, cross-section using
      Qutemol, 26

hydration surface representation, *135*, 138
importing files, *168*
methane, 30–31
organic molecule files, 33
parsers for, 31
PDB file format, 34–35, 36, 40–42
Peer communication, 13
Photography
knowledge of, 110
studio setup analogy, 113, 116, 149, 206
Photoreal rendering, *111*
Physical impact theory, 167
Physicochemical laws, 150
Physicochemical techniques, 23–26, 44
Pivot point, object, 60
Pixar, 99, *100–101*, 141
Plugins, 31
Polygon modeling, 44, 62, 63, 68, 70, 139
Portfolio of work, 206, 208
Post-production, 82, 179, 185, 201, 207
Primitives, 56, 57–61, 157
Principled shaders, 99, 139
Printed media, 91, 93
Pro forums, 7
Procedural modeling
animation, 171–174
growing an inorganic crystal, 174, *175, 176, 177, 178*
node system in Houdini, 174
paradigm of, 70–72
Programming languages, 70–71
Protein data bank *see* PDB (protein data bank)
Python, 71, 160

Q

Quick shading, 58
Qutemol, *24, 26*

R

Ray depths, 134, 143
Raytraced shadows, 79, 80

Real time simulations, 167
Red, green, blue (RGB), 88, 91
Red blood cells, 122–124
Red light, 85–87
References (scientific), research for modeling, 5, 6, 77–78, 205–206
Reflections, smooth/rough surfaces, *98*
Reflective material, 113, 116
Remeshing, 63
Render engines, 6
calculating your images, 141–146
coded laws of physics, 142, 143
separate passes, 183, 184–185, *189, 190*, 191
shortcuts, 142
Render passes, 183, 184–185, *189, 190*, 191
Render times, 146
Rendering
cell membrane example, *133*
default settings, 141
FX passes, 191
mathematical equation, 141–142
NPR, 198–200, *199, 200*
occlusion pass, 185, 191, *193–194*
Pixar definition, 141
saving file after, 92
settings, 119, *120, 127*, 142–146, 184–185, *184*
speed, 6, 87, 113, 116, 141
workflow, 115
Z-depth pass, 185, 191, *195, 196*
Rendering equation, 141–142
Revolving, 63
RGB (red, green, blue), 88, 91

S

Sample setup, 143
SAMSON, 33
SAS (solvent accessible surface), 21, 22, *23*
Scanning electron microscope (SEM), 33, 203–204
Scanning tunneling microscope (STM), 10
Scene setup, 75–96
classic studio, 116, *117, 118–119*

color and the viewer, 94–95
colors, 88–95
colors and mood, 85–87
common problems, 75–76
compositing your image, 93
emphasis only on part, 81–82
media selection, 91–92
scale in representation, 75–78, *76*, *77*, *86–87*
setting up lights, 79–88
Scientific concepts, metaphors and tools for explaining, 27
Scientific journals, 10, 91
Scientific references, research for modeling, 5, 6, 77–78, 205–206
Scientific software, 3, 155
Scitable (website), 157
Scripting language, 156, 160
Scripts, 7, 70–71, 72, 150–153
Self-illuminating channel, fresnel gradient, 113–114
SEM (scanning electron microscope), 33, 203–204
Serotonin molecule, pastel painting, *199*, 200, *200*
SES (solvent excluded surfaces), 21, 22, *23*
SFM (space filling models), 15, 21
Shaders, 98
    cell membrane example, *132*, 134
    choosing, 110–113
    and colors, 110
    fresnel gradient, 128
    NPR rendering, 198
    objects self-shadows, 142
    parameters, 99, *100–110*, 113–114, *126*, 138, 139
    Pixar, 99, *100–101*
    render setting, 119, *120*
    specular amount, *101–104*
    SSS, 128, *144–145*, 179, 188
    subsurface scattering, *107–110*
    translucency effects, *104–107*
    wet appearance, *166*
    workflow, 5–6, 115

Shading
    constant, 58
    quick, 58
Shadows, 79–80, *80*, 83, *84*, 128
Silhouette, 198, *199*
Skill levels, 207–208
Sky light, 81
Skydome light, 191, 198
Snapping, 60, 70
Social platforms, 1–2, 7, 208
Software
    2D painting, 198–200
    atom bonding rules, 18
    choosing, 4–6, 205, 207–208
    open source, 4
    parsers for PDB files, 31
    plugins, 31–32
    scripts, 7, 70–71, 72
    skill in specific software, 4
    used by author, 8
Solid structures, primitives, 61
Solidangle Arnold standard library, 116
Solvent accessible surface (SAS), 21, 22, *23*
Solvent excluded surfaces (SES), 21, 22, *23*
Solvent molecule, 22
Space, translate and rotate in, 60
Space filling models (SFM), 15, 21
Spatial volume, 21
Spheres, 15, 20, 29
Spline, 63, 69, 70, *172*
Spotlights, 82
State matrix, 60
Statistical thermodynamics, 167
STM (scanning tunneling microscope), 10
Stop-motion animation, Chimera, 150
Storyboards, 5, 178
Students
    common questions, 205–207
    dichotomy among, 2
Subdivision, 62, 70
Subsurface scattering, *107–110*, 113
Surface representation, *21*, 22, 23–27, 76

light scattering, *100–101*
  subsurface, *107–110*, 113
Synthetic images, 13

T

Textbook visualization, 10
Texture coordinates, 69
Texture mapping, 138
Transformation matrix, 60
Translucency effects, and lighting,
  104–107, 113
Turntable animation, DNA fragment, 157,
  158–160, *161, 162, 163*

U

Uncompressed image files, 6, 92, 200–201
UV coordinates, 141
UV mapping, 134–139

V

Van Der Waals surfaces (VDW), *20*, 21, *21*,
  23, 24
Vectorial formats, 92–93
Vertex(es), 57
  modifiers, 63–64, *65–66*
Vesta, 5, *19*, 32–33, 36–38
Viewer's attention, 75, 88, 94, 198
Virtual camera, 50–55, 114
  animation, 179
  depth of field (DOF), *52–53*, 185, 191,
    195, 196
  effect of using projection, *51*
  film gate, *54*
  focal length, *49–50*
  lighting, *131*
  number of iris blades, *55*
  settings, *166*
  white balance, 116
Virtual knife, 62
Virtual reality, 10–11
Virus, representations, 39–42, 79–88, *80*,
  83, 85, 86–87

Virus cross-section, *25, 26*
Visual appeal, 14
Visual metaphors, 13, 110
Visualization tools, audience specific, *14*
VMD, 5, 33
Volumetric lights, 82, 87–88
Volumetric shadows, 128

W

Water, as ubiquitous solvent, 88
Watercolors, 198, 200
WEHI Institute, YouTube videos, 14
Welding, 64
Wellbeing, 209
Wireframes, 58, 60, *79*, 114, *117, 122*,
  129–130
Wires, 19, *20, 21*
Workflow
  3D software and scientific software, 3
  authors, 5–6, 114–121, *121*
  color calibration, 93
  lighting and shades, 5–6
  material description, 113–114
  paper and pencil, 5, 6, 27
  plugins, 32
  printed media, 93
  scripts, 70–71
Work-life balance, 208–209

Y

YouTube videos, WEHI Institute, 14

Z

Zbrush, 4, 5
Z-depth pass, 185, *195*, 196
Zika virus, representations, 79–88, *80, 83*,
  85, 86–87
Z-matrix, 30
Zoetrope, 149, 183
Z-stack, 33

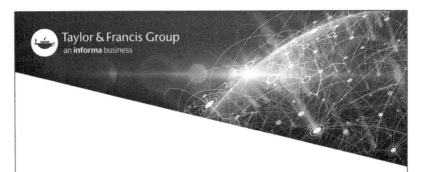